Benedikt Jager, Steffi Hobuß (eds.)
(Post)Colonial Histories –
Trauma, Memory and Reconciliation
in the Context of the Angolan Civil War

[transcript]

Bibliographic information published by the Deutsche Nationalbibliothek
The Deutsche Nationalbibliothek lists this publication in the Deutsche Nationalbibliografie; detailed bibliographic data are available in the Internet at http://dnb.d-nb.de

© **2017 transcript Verlag, Bielefeld**

All rights reserved. No part of this book may be reprinted or reproduced or utilized in any form or by any electronic, mechanical, or other means, now known or hereafter invented, including photocopying and recording, or in any information storage or retrieval system, without permission in writing from the publisher.

Cover concept: Kordula Röckenhaus, Bielefeld
Cover illustration: Film still from "My Heart of Darkness", Marius van Niekerk
Printed and bound in Great Britain by Marston Book Services Ltd, Oxfordshire
Print-ISBN 978-3-8376-3479-2
PDF-ISBN 978-3-8394-3479-6

Benedikt Jager, Steffi Hobuß (eds.)
(Post)Colonial Histories – Trauma, Memory and Reconciliation
in the Context of the Angolan Civil War

Postcolonial Studies | Volume 26

Contents

Black and White Dogs – Conceptual Encounters
Benedikt Jager (University of Stavanger),
Steffi Hobuß (Leuphana University Lüneburg) | 7

Angola
A Brief Historical Background
Ketil Fred Hansen (University of Stavanger) | 23

"They said we have to forgive each other"
Memory, 'Transitional Justice', and (Post)colonialism
in the Context of the International Screenings
of the Documentary *My Heart of Darkness*
Kaya de Wolff (University of Tübingen) | 37

The Ethics of Memory in *My Heart of Darkness*
Alexandre Dessingué (University of Stavanger) | 81

**Memory, Contradictions and Resignification
of Colonial Imagery in *My Heart of Darkness***
Steffi Hobuß (Leuphana University Lüneburg) | 99

Performing History
My Heart of Darkness from a Dramatist Perspective
Ketil Knutsen (University of Stavanger) | 119

**The Role of Music in Memorial Production
and Discourse in *My Heart of Darkness***
David-Alexandre Wagner (University of Stavanger),
Jon Skarpeid (University of Stavanger) | 145

"I don't trust in pictures"
Forms for Authentication in *My Heart of Darkness*
and Annekatrin Hendel's *Vaterlandsverräter*
Benedikt Jager (University of Stavanger) | 179

Memory, Trauma and Empathy
On the (Un)representability of the Civil War in Art
Nadine Siegert (Bayreuth University) | 207

***Miss Landmine* in Angola**
Negotiating the Political Aesthetics of the Mutilated Body
Nora Simonhjell (University of Stavanger) | 235

Notes on Contributors | 255

Black and White Dogs – Conceptual Encounters

BENEDIKT JAGER, STEFFI HOBUß

In 2016, the German illustrator and author Birgit Weyhe published a remarkable graphic novel entitled *Madgermanes*. The book focuses on the experiences of a particular group of people, namely migrant workers in the GDR. During the Cold War, communist countries 'invited' workers from socialist countries in the Third World. For example, over 20, 000 workers from Mozambique came to East Germany to learn from the technologically more developed sister state. Their story has been one of disappointments. The assumed learning opportunity quickly developed into hard and underpaid work exploitation. Instead of the official socialist solidarity, these workers encountered blatant daily racism. The economic benefits were stolen by corrupt politicians in the home countries. They became

"Madgermanes", the mad Germans who were 'made in Germany' and who neither fitted into the reunified Germany nor into Mozambique (Weyhe 2016).

Many of the questions discussed in Weyhe's book are also crucial for the publication at hand. The film *My Heart of Darkness* also leads us back to the time of the Cold War. After the liberation of Angola from the Portuguese colonial power, a civil war emerged involving socialist and capitalist countries, such as Cuba or South Africa. Marius Van Niekerk came to Angola as a white South African paratrooper and was traumatized by his experiences and deeds in this war. He was driven into exile and moved to Sweden, but was haunted by the past even there.[1] Like the foreign workers in the GDR, he was stuck in an in-between that complicated and troubled his identity. As Birgit Weyhe puts it (as can be seen in the opening illustration), the question of memory and the strange ways of remembering are crucial for both works: The dog in heat is straying around and is following the traces but, as the illustration indicates, the traces are meshed up in a chaotic knot. Van Niekerk moved from the South to the North, but carried with him the memories and the scars from the war, a war that was fought in the spirit of Apartheid. Van Niekerk's project, which is documented in the film, is to gain sanity and to overcome madness, not only for himself, but also for his family. Fighting his own ghosts from the war, he is consciously trying not to transfer his trauma onto the generation of his children. As Marianne Hirsch (1997) put it in coining the term "postmemory", a remote conflict is inflicted upon this generation. The questions of memory and the transmission of memories are therefore central in *My Heart of Darkness* and in *Madgermanes*.

The straying dog behaves like the memories of the traumatic experiences. It appears unexpectedly at places, it is unreliable, and it cannot be controlled deliberately or by rational means. Thus, it shares one of the features of "travelling memory" (Erll 2011). Other aspects of this travelling memory in the case of *My Heart of Darkness* are the processes of remediation, since it is a documentary film that alludes not only to Joseph Conrad's novel *Heart of Darkness* but also to Francis Ford Coppola's *Apocalypse Now*, the transnational context of the Angolan Civil War, Van Niekerk's life and, finally, the context of recent memory studies.

---

1   See also Marius Van Niekerk and Peter Tucker (2007).

In this sense, there are striking parallels between the three projects, including our research. Both the graphic novel, the film and our research have their point of departure in a white perspective on Africa. Both Birgit Weyhe and Marius Van Niekerk are biographically connected to Africa. Weyhe spent her childhood in Uganda and Kenya and reflects on this in her book as motivating her sensibility for the questions of migration and in-between-ness, thus also explaining how she came upon the story of the foreign workers from Mozambique in the GDR in the first place, a part of history neglected in the German public after the unification in 1990. In research on modern autobiographical writing, the phenomenon of the "story of the story" has been developed over the last decades. For John Paul Eakin this phenomenon has been an important sign marking the turn of the autobiography away from the celebration of the autonomous (white, male) ingenious subject to an acknowledgment of the interweaving of our lives with those of others, towards relational autobiographies (Eakin 1999, 56). Following this idea, we would like to say a few words about the story of our story, how we got involved with the Angolan civil war as seen through the lens of a South African living in Sweden.

This publication is the result of an international collaborative research project. In December 2014, we held a workshop at the University of Stavanger. We were a group of researchers from Sweden (Karlstad), Germany (Lüneburg), Great Britain (London), and Norway (Stavanger). The group had been established as a research network in memory studies since 2011, led by Alexandre Dessingué (University of Stavanger) and funded by the Norwegian Research Council through the "Program for the Cultural Conditions Underlying Social Change" (SAMKUL). The main objective of this project was to discuss and analyse the new challenges in the uses of cultural memories and the (de-)construction of national myths in a global age.[2] The last workshop of this project organised at the University of Stavanger focused on uses of memory theories and methodologies in a comparative and interdisciplinary approach, and in light of a common case study, i.e. the documentary film *My Heart of Darkness*.

---

2   At the same time, many of the participants were members of the COST network In Search for Transcultural Memory in Europe (ISTME), funded by the EU COST program in 2013-2016.

For some of the researchers, it was their first visit to the Southwest of Norway and, being only two weeks before Christmas, they were not disappointed. The town of Stavanger and the landscape around was covered by a thick layer of snow. The campus of the University was bathing in pure whiteness and the image was completed by deer passing by outside the classroom, in which we were gathered. This Disney-like image is of course too good to be true. Our hospitable and generous host, namely the University of Stavanger that has supported this publication in many ways, is the result of blackness. Stavanger is the capital of the Norwegian Oil industry and has therefore become one of the most expensive towns in the world, especially bearing in mind that it has 125 000 inhabitants. The use of fossil fuels is the acknowledged reason for the increasing climate change and the domestic participants had a hard time shattering the image of Stavanger as a winter wonderland the visitors had acquired. Normally, the winter in Stavanger is grey, with lots of rain, no snow and three to four degrees above zero, and to be honest it was like that even before the climate changes. The children here are not born with skis on their feet.

Why this report about images of the North and the locatedness[3] of a research project? The 'Madgermanes' from Mozambique experience several forms of xenophobia in the North and one character answers this challenge with an African proverb (Weyhe 2016, 62):

---

3 For the concept of locatedness as a response to the transnational and transcultural turn in the context of memory studies, see Radstone (2011).

This little strip in some way captures many of the guidelines for our research, as well as the challenges we had to meet. All the contributors of the papers in this volume are white academics dealing with Angola, a country that has not been in the center of public interest in the West. Thus the effort we try to put into these collected essays can be understood as part of the project of critical race studies, or *critical studies of whiteness*. However, since the last decade, *critical studies of whiteness* have been met with a considerable amount of critique. White academics who try to be aware of their whiteness when dealing with the topics discussed here still tend to reproduce white supremacy and colonial and racist power relations. There is no way out of this problem. We are confronted with a dilemma that is one of the fundamental problems of neocolonialism: Either we try to give a critical analysis of the constellation we find in *My Heart of Darkness*: White directors making a documentary about the Angolan civil war and the traumata of the war. However, we have to reproduce and repeat racist categories and relations by simply speaking as white academics about this constellation. There is no possibility of escaping the "reiterations" (Derrida 1988) of white supremacy and biased categories. Alternatively, trying to avoid such repetitions and references to universalism of mankind means risking color blindness (cf. Wollrad 2005): "Even though it has become commonplace to utter rote phrases such as 'race is a construct' or 'race does not exist', etc., race itself shows no evidence of disappearing or evaporating in relevance." (Crooks 2000, 4) There is no escape from this dilemma and it can only dealt with through an ongoing process of discussion. In this troubled situation, it is always necessary to speak against oneself, or to use language against its own notions and restrictions.

We understand our contributions as one step in this ongoing process and hope for further discussions and dialogue. In any case, our readings of the documentary in the context of memory politics, memory culture and the ethics of memory try to show the importance of going beyond simple binaries and being aware of ambivalences and contradictions. In the context of memory studies, simple binaries always seem to be dead ends, as Astrid Erll states when referring to the binary of history vs. memory (cf. Erll 2008, 7).

With this in mind, let us have a look at the leopard anew. The leopard in the first image has white stains that change into black stains when zooming in on the animal's skin. The background is brushed with brown strokes introducing a third color, transcending the dichotomy of black and white. It is our hope that the articles collected here can contest the oppositional thinking in exclusive categories.

For some of the researchers, these topics are not only of academic interest, but are also part of their everyday lives. Being migrant workers from France and Germany living with their families in a cross-cultural context in Norway actualizes many of the problems (and chances) that migration and exile raise. Of course, our experiences are much more 'pleasant' and less traumatizing than what *My Heart of Darkness* deals with. What we have in common is the awareness of the journey of memories from one context to another.

Another fundamental difficulty is the factor of gender: *My Heart of Darkness* is a male-dominated documentary. Some gender aspects that are important in the Angolan context are dealt with in this volume, e.g. the male-dominated agent and agency of the documentary in opposition to the objectification of women, Van Niekerk's relationship to his daughters, all the veteran's relationships to their (former) wives, the fact that until today the victims of landmines have been mostly women and children, and the role of gender and the female body in artists' practices. A systematic analysis of the gender constructions and gender relations in *My Heart of Darkness* is still a desideratum in research, especially since memory studies sometimes tend to marginalize women's memory work.

The image we chose for the book cover is used in the documentary (42'-45'), when a testimony of a horrible event is reported, namely the massacre of burning innocent women by the troupes of Savimbi. This is reminiscent of the many women and children who were victims of the civil war, but who are most often marginalized when it comes to collective memory practice.

Blackness and whiteness, men and women, neocolonialism and north-south relations, adults and children are all categories that are inseparably intertwined in the memories and narratives of the war. The *intersectionality* of these aspects, in terms of the notion Kimberlé W. Crenshaw coined to describe the overlapping or intersecting social identities and related systems of domination, oppression, or discrimination (Crenshaw 1989), can be seen to be at work in *My Heart of Darkness*. During the different scenes, different categories, such as race, gender, age, and others, become more or less important or dominant in relation to others.

In sum, our endeavor documented in this volume, as well as the documentary *My Heart of darkness* as a cultural artifact itself, can be considered as highly ambivalent concerning the negotiations of the Angolan Civil War, the traumatic experiences, and all the related topics, such as (post-) colonialism, racism, critical race studies, and gender. Each of the contributions deals in some way with some of these difficulties, ambivalences and contradictions.

The initial article by Ketil Fred **Hansen** opens our ensemble of papers by unfolding fundamental information about the Angolan civil war in order to make it easier to understand the references and dialogues in *My Heart of Darkness*. While focusing on the period from the Portuguese surrender up to the year the film was shot (2007), his chapter gives an overview and also includes a short description of central historical developments prior to 1975 when these are considered relevant to the civil war.

As Kaya **de Wolff** points out, the controversial memory discourse around *My Heart of Darkness* speaks both of Western domination of the reconciliation paradigm and an elite-driven top-down-memory politics, as well as of local narratives from below and a popular counter-memory. De Wolff focuses on the controversial discourses surrounding the screenings of the documentary in different contexts. In her paper, "'They said we have to forgive each other': Memory, 'Transitional Justice', and (post)colonialism in the context of the international screenings of the documentary *My Heart of Darkness*", she points to the different interpretations that have been given to different contexts of the screening. Based upon the idea that the documentary film as a cultural text unfolds its meaning in interactions with an active audience, she unfolds three different readings of the documentary. According to the first reading, *My Heart of Darkness* is a universal film about veterans that portrays them as suffering from post-traumatic stress disorder (PTSD) and seeking for forgiveness, social rehabilitation and recognition. The second reading understands the film as a model for national reconciliation in Angola and raises a critique of top-down memory politics. According to the third interpretation, the documentary gives a critique of the exploitation of the people who were pressured into the civil war in Angola and left on their own in its aftermath. In this interpretation, the official memory discourse of national reconciliation neglects and silences the vernacular memory of the Angolan people and, most of all, the Angolan veterans.

After this detailed case study and analysis of the reception and the discursive adaptation of the documentary in its different contexts, the following papers turn towards the interpretation of its contents and its aesthetics. In his contribution "From common to shared memory and from forgiving to forgetting", Alexandre **Dessingué** deals with the ambivalences of the documentary itself. He asks to what extent it actually works as a process of forgiveness and reconciliation, as this intention is clearly introduced in the trailer. Starting from the observation that the documentary deals with individual acts of remembrance, as well as with collective acts of re-mediation and forgetting, Dessingué analyses the film's dialogues in order to show that the intention is much more ambitious than explicitly introduced. He reads the documentary as a "construction of a shared experience", referring to the philosophy and ethics of memory like Arendt's notion of the "banality of evil", and using Margalit's concepts of "common memory" and "shared memory" (cf. Margalit 2004). While a common memory is an aggregate of individual acts and keeps a polyphonic nature, a shared memory, according to Margalit can be seen as a result of a shared and calibrated cultural representation. However, *My Heart of Darkness* remains a mere attempt to change an individual journey into a collective one, as is revealed by a closer look at the kind of forgiveness the documentary is about. Dessingué differentiates between forgiveness in the sense of a "covering-up model" and a "blotting-out model", again according to Margalit. While the covering-up model integrates an acceptance by the victims of forgetting, no clear victim-perpetrator-relationship can be found in the documentary. Rather, it uses a blotting-out model of forgiving, because it mostly shows an authoritative initiative of the main character Marius Van Niekerk. Thus, the ethical-philosophical focus on different types of memory and different types of forgiveness highlights the ambivalences of the documentary.

In Dessingué's interpretation, Van Niekerk tries to establish an "individual, active and voluntary" overview of past events – what has proved to be particularly problematic. Steffi **Hobuß** deals with a similar question and looks at the contradictions that occur because the documentary uses examples of colonial imagery in order to change the meaning of the depicted journey from an individual into a collective one. Against the backdrop of an introduction to the idea of collective memory, understood in terms of present memory acts rather than as a representation of past events, her article about "Memory, Contradictions and Resignifications of Colonial Imagery in *My Heart of Darkness*" examines contradictions as a central part of the film's aesthetics and provides some examples of resignification and remediation of colonialist and racist imagery. Through the aesthetic means of the documentary, it shows more than the expressed individual intention of the main character, which can be understood as Van Niekerk's wish for private ownership of his memories, a desire that is impossible to fulfill. The issues of colonialism and racism continuously present throughout the documentary, through the allusions to Joseph Conrad's *Heart of Darkness* and Coppola's *Apocalypse Now,* seem to be swallowed up by the "universal" aims of forgiveness and recon-ciliation. Other aspects include the use of daylight, firelight and darkness, and the pictures of animals and nature are by chains of resignification connected with colonial imagery.

The understanding of memory as present acts, rather than depictions of the past developed as a theoretical framework by Hobuß also forms the subject of Ketil **Knudsen**'s contribution about "Performing History – *My Heart of Darkness* from a dramatist perspective". Knudsen focuses upon *My Heart of Darkness* from a dramatist perspective. He uses the "pentad analysis" according to Kenneth Burke to elucidate the motivational forces that drive the film. This methodological framework allows seeing the film as performative, determined by meaning practices that relate the past to the present and the future. As in John L. Austin's *Theory of speech acts*, from the perspective of pentad analysis, language is a mode of action rather than a mode of knowledge, or describing the world by reference. Moreover the film can also be analyzed as an act rather than a representation of something. Knudsen considers regaining dignity and purging the present by getting rid of the memories from the war as the act carried out during the narration. For Van Niekerk, the purpose of the act is to redefine his identity by purgatory and to become a father again. Knudsen views Van Niekerk and the other veterans as the main agents. Although all the main characters were at least perpetrators and victims during the war, as agents they are shown only as victims. While there are other persons present in the film, such as the Angolan villagers or family members, objectification rules them out as agents. By asking for the aspect dominating the film's narration, it is clearly the agent who dominates. All other aspects are related to and subjected to the agent. Van Niekerk's aim of healing himself requires constructing himself and the other veterans as victims and at the same time using the others as a means for his individual project. On the whole, in the light of Knudsen's investigation, *My Heart of Darkness* is driven by this individual project rather than by an interest in the scene, i.e. the Angolan context. Although it is methodologically connected with approaches such as oral history, history from below and research on reconciliation, and explicitly deals with the question of how we can come to terms with history today, it reproduces domination to some extent.

In "The Role of Music in Memorial Production and Discourse in *My Heart of Darkness*", David **Wagner** and Jon **Skarpeid** focus on the interplay of the film's narration with the film's music. In a first step, they discuss the role of music in feature films and documentaries in general and link this discussion to the seldom-debated question of music and memory. In the everlasting dispute about music and reference, they argue that music in connection with other codes of the film contributes to and deepens the emotional and semantical impact of film. In accordance with Annabel J. Cohen's framework of internal and external semantics of music at different levels (congruence and association), Wagner and Skarpeid present the possibilities inherent in the choice of tone, dynamics, timbre, and rhythm and tempo.

In some way, the film *My Heart of Darkness* is an untypical documentary because it uses extradiegetic film music in an extensive way. Over half of the film is accompanied by music by the Swedish composer Jan Anderson and breaks with the journalistic rationalism and observational minimalism that almost has to be seen as a dogma in documentaries (the same applies to Annekathrin Hendel's *Vaterlandsverräter* that Jager uses in his reading as a contrasting counterpart to Van Niekerk's film.) The distribution of the parts with music and without music are also of great importance. On a semantic level, Wagner and Skarpeid observe a polyphonic issue in the discussion of the four former war veterans. This choir of voices is divided in relation to the film music: only the comments by the narrator Marius Van Niekerk are accompanied by music. This fact offers the audience a greater surface for identification with his project and accentuates his priority. In painstaking analyses, Wagner and Skarpeid show how the use of musical motives contributes to deepening the emotional impact of the film's narrative. Most interesting in this way is their reading of the "The Milgram Effect", the music accompanying the end credits. The music in this part reintroduces an ambivalence from the margins of the film and problematizes the optimistic solution provided by the first-person-narrator in the cleansing ceremony. While this is a very subtle comment, it finds an equivalent in a very short scene when the reconciliation is reached, the goat is sacrificed, and the former enemies are friends. This peace is broken by Patrick and Samuel roleplaying (like children) their respective part in the battle of Cuito Cuanavale. This scene questions the equilibrium that the film has reached and implements a new polyphony, possibly according to the doubled authorship of the film.

The use of music can also be seen as a mode to strengthen the impact of the film, to increase its ethos. Benedikt **Jager** starts his article "'I don't trust in pictures': Forms of Authentication in *My Heart of Dar*kness and Annekatrin Hendel's *Vaterlandsverräter*." with a similar question. In Van Niekerk's film, the reliability of documents is central and it is of great importance for Marius that his fellow travellers accept these as authentic. The article is interested in which strategies the film uses to create an aura of authenticity. The term "authenticity" is one of the key terms of the second half of the 20th century and is in part a symptom and also in part a solution of the crisis of modernity. As the theoretical discussion of the term has shown, it functions as a promise to overcome 'the transcendental homelessness' (Lukács 1971, 32+52), a promise that it ultimately cannot deliver. In this way, Jager's approach does not focus on the question of whether the films are authentic, but rather on whether they create an aura of authenticity. The starting point is the observation that both films share some motives in telling their stories. On a macro level, the motive of the boat-travel and the campfire as a place for storytelling is used in both narrations. In *My Heart of Darkness* the setting is loaded with implications that signal that the sphere of corrupt culture will be transcended. Travelling up the river is a going back to nature, to archaic forms of life that function as a rebirth for the traumatized soldier. The peak of this development is the cleansing ceremony and the sacrifice of a goat. A closer look at this suggestive story shows that this the staging of authenticity is more a tangle of contradictions than a sole Ariadne's thread. The discussions of the status of Van Niekerk's war-pictures are central here, but so are questions of genre.

Annekatrin Hendel's films often deal with the question of authenticity. As a former citizen of the GDR, she is fascinated by the role of the intelligence (Stasi) in society and the cultural sphere. By scrutinizing this field, she ends up in indecisive twilight zones, that have consequences for the staging of her films. While *My Heart of Darkness* tries to persuade the audience that reconciliation and salvation is possible by going back to nature, *Vaterlandsverräter* shows the prevailing aporias and makes them visible by different strategies. The status of documents is discussed and remains disputable, and the importance of genre as a framing factor for our understanding is not suppressed but marked. In using (fictional) oil paintings as illustrations that are rooted in conventions and that allude to trivial genres she finds a strategy to show how these events are part of a construction, one that is at last arbitrary and not rooted in the authentic.

In the last part of the book, we present two contributions that examine the Angolan Civil War and its aftermath in relation to other art projects. Nadine **Siegert** deals with Fernando Alvim's project *Memorias – Intimas – Marcas* and Jo Ractliffe's photographic series *As terras no Fim do mundo*. Both these projects share, together with Van Niekerk's film, an interest in important places of the Angolan war: the battlefields of Cassinga and of Cuito Canavale. The latter gained almost mythological status as the African Stalingrad (cf. Sanney 2006). The Angolan Alvim and the South African Ractliffe also travel back to these two places where war atrocities took place in 1978 and 1987-1988. Siegert is most interested in discussing the different strategies by approaching the traumatic events and connects this to the possibility of representation of trauma. Using the works of Aleida Assmann, Dominick LaCapra and, most importantly, Jill Bennett, she traces different strategies of memorization in the artworks. The watershed of her argumentation can be found in the opposition of representability and un-representability of trauma. Her arguments question whether the different approaches to representability lead to different concepts of politics of memory and reconciliation. While Van Niekerk opts for a representation of traumatic events as something that can be left behind, Alvim and Ractliffe rely on other strategies. Against this background, Siegert discusses a question similar to Jager's. Authenticity and representation, if not twins, are at least close siblings. As opposed to Van Niekerk, Ractliffe and Alvim are preoccupied with the locations of the war, not in order to reconstruct a historical truth, but to engage actively with trauma in a communicative way. The communication is not governed by knowledge and truth, but by relations and negotiations between the participants of the projects. While the photographs in the shoebox are undisputable in Van Niekerk's understanding, the other projects have silence, emptiness and a lack of outspoken signs as a starting point. Against the backdrop that these projects are performative, they open for a concept of participation and affectual experience. They address the process of working through a traumatic past, not the result, which leads Siegert to interesting observations about the relationship of personal memory to collective memory processes.

Nora **Simonhjell**'s "Miss Landmine in Angola. Negotiating the political aesthetics of the mutilated body" also deals with the relationship of the individual and the collective. Angola, alongside Cambodia and Afghanistan, is the country in the world that is still suffering the most from landmines placed during the civil war. The stories emerging from these weapons are different from the stories told by the war veterans in Van Niekerk's film since the victims of landmines are mostly women and children. Most narrations about the war are told by male soldiers and the gendered perspective is often neglected in the collective memory processes. An illuminating example from a German context is that the common writers of diaries are mostly women, while only a third of diary keepers are men, but this composition is reversed in wartime. The Deutsches Tagebuch Archiv in Emmendingen shows that the private narrations about the Second World War are clearly male-dominated. In Norway, we can observe the same phenomenon in relation to the resistance narrative: the stories of female members of the resistance are concealed. The Norwegian artist Morten Traavik worked with these issues by changing the focus of one of the most male perspective's on the female body: the beauty contest. In 2008, he arranged a beauty contest for Angolan women who had suffered severe injuries from landmines.

Simonhjell's perspective is twofold: She reads the project in the frame of the genre beauty pageant and relates these findings with insights from the field of disability studies. Through a short retrospective on the history of beauty pageants, research could show that the bodies of the female participants had to be understood as symbolic signs. The beauty queen had to be seen as an idealized ambassador for the community, a narcissistic reflection that suppressed all 'negativity'. This is the hub of Traavik's project. Simonhjell furthermore shows that he is not interested in incorporating these women in the logic of globalized capitalism, but in focusing on the mechanisms of displacement of the deviant. By showing the scars of the mutilated bodies, Traavik indirectly shows the history of the Angolan Civil War and the still ongoing struggle in everyday life, and tries to empower the affected people. They are not victims that need our compassion, but rather living women with dreams and challenges. In a broader context, the project aims to address questions of otherness and normality. The disabled can be addressed as the unruly body that challenges our categorization of entity, harmony, or the dualism between ill and well. In this way Traavik's ironic inversion of the slogan of the *Miss World contest*, "Beauty with a purpose", into "Beauty with a difference" is striking.

The critical reader may ask whether a whole book about a film that has not gained international fame is perhaps giving it unwarranted attention. The editors and contributors hope that the articles compiled in this edition both have heuristic value with regard to the film and also discuss broader questions. This introduction started with the image of memory as a dog in heat. With some modifications, it is also suitable for the situation of the researcher – we like to see orselves as sleuths in the humanities, in society. This project has in many ways been a knot, a crossing of traces that we tried to follow, to untangle and, finally, to tame. If one takes a close look at Weyhe's frame of the dog of memory, one can distinguish several layers. The white brushings are made by hand and fingerprints are recognizable. In the same way, this project on a remote land and a conflict almost forgotten in Norway and Germany became personal in an unexpected way. At a gathering at the Academy of Science in Oslo, the editors met a well-renowned scholar from Oxford. In the normal coffee-break-chat about ongoing research, we mentioned the Angola project and evoked a reaction other than the expected courtesy. In a novel of the 19th century, it would be called: "He turned pale!" And he told us: "That could be my story – that is, why I went from my homeland South Africa into exile to Great Britain!" The ways of the dog, of memory and research, are unpredictable. In this way, we shake the water from our fur, just like a dog on the shore, and deliver this ball of research to whom it may concern. Perhaps somebody else would like to throw the ball ahead to other stray dogs on the beach.

Stavanger and Lüneburg, April 2017

REFERENCES

Crenshaw, Kimberlé (1989). "Demarginalizing the Intersection of Race and Sex: A Black Feminist Critique of Antidiscrimination Doctrine, Feminist Theory and Antiracist Politics". *University of Chicago Legal Forum.* 140, pp. 139-167.

Crooks, Seshadri (2000). *Desiring Whiteness. A Lacanian Analysis of Race.* New York: Routledge.

Derrida, Jacques (1988). "Signature Event Context", in: J. Derrida, *Limited Inc,* Evanston, Illinois: Northwestern University Press, pp. 1-23.

Eakin, Paul John (1999). *How our lives become stories: Making selves.* Ithaca: Cornell University Press.

Erll, Astrid (2008). "Cultural memory Studies. An Introduction". In: Astrid Erll and Ansgar Nünning (eds.), *Cultural Memory Studies: An International and Interdisciplinary Handbook.* Berlin: de Gruyter, 1-18.

Erll, Astrid and Stephanie Wodianka (2008). *Film und kulturelle Erinnerung. Plurimediale Konstellationen.* Berlin: de Gruyter.

Erll, Astrid (2011). "Travelling Memory". In: *Parallax,* Volume 17, Issue 4: Transcultural Memory, pp. 4-18.

Hirsch, Marianne (1997). *Family Frames: Photography, Narrative, and Postmemory.* Cambridge, Mass.: Harvard University Press.

Lukács, Georg (1971). *Die Theorie des Romans. Ein geschichtsphilosophischer Versuch über die Formen der großen Epik.* Neuwied: Luchterhand.

Margalit, Avishai (2004). *The Ethics of Memory.* Harvard University Press.

Radstone, Susannah (2011). "What Place Is This? Transcultural Memory and the Locations of Memory Studies". In: *Parallax,* Volume 17, Issue 4: Transcultural Memory, 109-123.

Saney, Issac (2006) "African Stalingrad. The Cuban Revolution, Internationalism, and the End of Apartheid". In: *Latin American Perspectives,* Issue 150, Vol. 33, No. 5, September 2006, pp. 81-117, online http://www.normangirvan.info/wp-content/uploads/2009/01/saney-african-stalingrad

Van Niekerk, Marius and Peter Tucker (2007). *Behind the Lines. Healing the Mental Scars of War. The Story of a South African Paratrooper.* N.p.

Weyhe, Birgit (2016). *Madgermanes.* Berlin: avant-verlag.

Wollrad, Eske (2005). *Weißsein im Widerspruch*. Königstein: Ulrike Helmer Verlag.

Angola

A Brief Historical Background

KETIL FRED HANSEN

In the film *My Heart of Darkness*, four former combatants in the Angolan civil war (1975-2002) meet to heal their traumas, some years after the end of the war. All of them were hired as soldiers or guerrilla fighters between the ages of 14 and 17. Which movement they were recruited to seems more or less coincidental. Marius van Niekerk is a white ex South African Defense Force (SADF) parachutist and the filmmaker. Patrick Johannes was coerced to fight for years for the MPLA regime's FAPLA. Samuel (Sammy) Machado Amaru was recruited by force into the UNITA guerilla group. Finally, the oldest of them, Mario Muhonga, who fought for Portugal against all the others (MPLA, UNITA and SADF) before Angolan Independence, and then, after Independence, was recruited by the South Africa Defense Force to fight against the MPLA regime.

To make it easier to understand the references and dialogues in the film, this chapter seeks to situate the historical context of the war. While focusing on the period from the Portuguese surrender up to the year the film was shot (2007), this brief overview also includes a short description of central historical developments prior to 1975 when these are considered relevant to the civil war.[1]

---

1 This chapter is based on the literature mentioned at the end of this chapter. In addition, I have used numerous reports and policy briefings from the Institute for Strategic Studies in Pretoria, the Angola country chapters in Africa Yearbook (Leiden: Brill), and the Fellesrådets Afrika årbøker (Oslo) published regularly during the last 10-20 years. News articles from the New York Times

PORTUGUESE PRESENCE IN ANGOLA

Portuguese merchants had been present on the Angolan coastline since the seafarer and explorer Diogo Cão "discovered" the land in the late 14th century. A few hundred Portuguese migrants and soldiers installed themselves in Luanda and fortified the town around 1575. Initially attracted by gold, capturing and trading slaves became the most important activity for the Portuguese in Angola from mid-1500 to late 1800. During that period, some three million people were deported as slaves from the area, most of them to Brazil.

Portugal formally colonized Angola only after the European partitioning of Africa negotiated during the Berlin-conference in 1884/85, its exact frontiers were negotiated for another 6 years, and it took the Portuguese state more than thirty years to gain virtual control over the population in their new colony. Using harsh methods including torture and flogging, public executions and imprisonments, threats and forced labor, the Portuguese slowly began to master the territory.

Settlers from Portugal arrived to clear land and cultivate cash crops. By the 1830s, they had established numerous coffee plantations and cotton production started in 1926. Both coffee and cotton production rose steadily, making Angola one of the world's largest producers during the 1960s.

After the first Portuguese discovery of diamonds in 1912, fortune hunters arrived to search for gold and diamonds. By 1917, together with British, Belgian and US interests, the colonial state established its first diamond company, *Diamang*.[2] Offshore oil production in Angola started in 1955, but it was not until 1973 that oil surpassed coffee as the country's single most valuable export. Coffee, cotton, diamonds and oil were thus the main economic resources in colonial Angola.

After the Second World War, colonial regimes worldwide started to face more organized resistance, both at home and in their colonies. All over

---

    and Forbes have been used to supplement when necessary. Rather than a research article, this is a background chapter, and I have deliberately avoided using direct references to literature in this article.

2   However, when the state dissolved the company in the late 1980s, it had contributed little to development due to heavy extraction investments, public corruption, and substantial diamond smuggling.

colonial Africa, western educated Africans established political parties. Some political parties sought only more governmental influence while others wanted total independence. British and French governments allocated more civil and political rights to (educated) Africans in their colonies during the 1950s, however, colonial opposition continued to gain momentum. Starting with the Sudan in 1956, all British colonies in Africa had gained their independence by the mid 1960s, and most French colonies in Africa were given their independence in August of 1960.

For the Portuguese colonies however, it was different. Portugal was among the poorest countries in Europe and was governed de facto by Prime Minister Antonio Salazar from 1932 to 1968. Salazar's regime was anti-communist, fascist-friendly and dictatorial. A secret police force supported by informers repressed political dissidents. Political elections were only a formality as Salazar personally chose every government minister. Political freedoms were few and censorship widely practiced. Until his death, Salazar was determined to retain the colonies to maintain Portugal's international grandeur and to contribute to its economic progress.

The Salazar regime governed Angola in a more dictatorial fashion than Portugal, depriving Angolans of most political rights and civil liberties. As British and French colonies gained independence, there were no signs of political liberation in Angola. In March 1961, a small incident – retarded payment of salaries to plantation coffee laborers in northern Angola – escalated into an armed conflict. The Portuguese managers at the plantation interpreted the coming of laborers towards their offices as an attack on them and opened fire. Angolans laborers, supported by members of the minor Bakongo liberation movement *Uniao das Populacoes de Angola* (UPA), a group headed by Holden Roberto, fired back. Over the coming three months of intense conflict, hundreds of whites and thousands of blacks were killed. The colonial state expelled some 250.000 Angolan and Congolese workers to the Congo. The already aggressive colonial regime turned more repressive than ever, arresting thousands and executing hundreds of potential political dissidents all over the country.

The colonial regime hardened both physically and economically. Salazar sent a huge conscripted army to Angola, creating frustration among soldiers and generals at the point of mutiny in Lisbon. Unmotivated Portuguese army officers in Angola quickly developed a taste for business and soon controlled the illicit currency exchange. They also participated in more-or-less legal export-import trade from Angola. While African dissidents were the main target for the colonial regime, white Portuguese also lived in fear for the secret police. However, if they remained politically calm and passively supported the regime, most Portuguese in Angola enjoyed a better living standard than their brethren in Portugal.

These developments encouraged Angolan fight for colonial liberation (1961-1975), a fight that continued as a civil war for state control after Independence (1975-2002). Three armed movements were central in both these wars: Neto's MPLA, Roberto's FNLA, and Savimbi's UNITA.

COLONIAL LIBERATION WAR (1961-1975)
MPLA UNDER NETO'S LEADERSHIP (1956-1979)

The People's Movement for the Liberation of Angola (*Movimento Popular de Libertação de Angola*: MPLA) was established in 1956 when two minor communist and anti-colonial parties merged. From the beginning, MPLA was dominated by educated, urban Africans in and around Luanda.

A well-known political dissident and left-winger, Augustinho Neto, was elected the MPLA's first president. At that time, he was still in prison in Lisbon where he was given a seven-year sentence for spreading communist ideas and agitating for Angolan liberation. Trained in Portugal as a medical doctor, he married a Portuguese and returned to Angola in 1959 to practice as a doctor[3]. Neto, a well renowned poet, was also politically outspoken, and a year later, the colonial regime arrested him for spreading communist ideas. When Amnesty International was established in 1961, Neto was selected as the organisation's first prisoner of conscience. When released from jail, Neto travelled to meet Fidel Castro in Cuba. Neto and Castro shared the same communist ideology and the MPLA gained military support from Cuba for years, ending only when the Cold War ended. The Soviet Union also supported the MPLA ideologically and military. In fact, the Soviet Union considered the MPLA so ideologically important that the they awarded Neto the Lenin Peace Price for 1975-1976 for his role as a

---

3   In 1960, Neto was one out of five African western educated doctors practicing in Angola.

communist colonial liberation fighter. Neto remained the MPLA's leader until his death in 1979.

FNLA UNDER HOLDEN ROBERTO'S LEADERSHIP (1961-2007)

In 1961, Holden Roberto's UPA, together with other minor political liberation groups based in northern Angola, created a common, anti-communist guerrilla movement, the National Liberation Front of Angola (*Frente Nacional de Libertação de Angola*: FNLA). The United States supported the FNLA with military equipment and cash in the years just prior to Independence hoping it would function as a counterweight to the communist MPLA. Due to Portugal's fascist friendly regime in Angola, Israel also gave military training and weapons to the FNLA prior to Independence. In addition, the Mobuto's regime in Zaïre (today Democratic Republic of Congo) generously supported the FNLA and parts of the civil war with cash, weapons, and access to its land. Mobuto and Zaïre's support to the FNLA intensified after Roberto married a relative of Mobuto's wife in the early 1970s, and by 1974 the FNLA was by all appearances the strongest liberation movement in Angola. However, in 1976, MPLA forces severely defeated the FNLA, and Roberto and his men fled to Zaïre. They continued to follow their political aspirations in diaspora, though with less ambitious guerrilla tactics and less foreign support. They did, however, succeed in kidnapping MPLA leaders and bringing them to Zaïre where some were brutally killed. Personally, Holden Roberto had been financially secured by the CIA from 1963 onwards, however, direct US military funding to the FNLA only took place for a few years just prior and after Independence. In fact, by the late 1970s and early 1980s most of the FNLA soldiers returned to civilian life without being formally demobilized. The remains of the FNLA transformed from a guerrilla movement to a political party just prior to the presidential elections in 1992. As the FNLA's candidate, Holden Roberto received only two percent of the casted ballots. Until his death in Luanda in 2007 at the age of 84, Roberto continued to preside over a fragmented and relatively unimportant FNLA.

THE ESTABLISHMENT OF THE UNITA UNDER THE LEADERSHIP OF JONAS SAVIMBI (1966-2002)

In 1966, after fighting for a short period together with the FNLA, Jonas Savimbi founded a third liberation movement in Angola, the National Union for the Total Independence of Angola (*União Nacional para a Independência Total de Angola*: UNITA). UNITA was dominated by members of Savimbi's own ethnic group, the Ovimbundu. Savimbi was trained in both Portugal, Switzerland and China. A true intellectual, he spoke seven languages fluently, including English, and he was very well read. In addition, he was a very charismatic leader. Despite this, UNITA had less than 1000 armed men in 1974 and was far less important than both the MPLA and the FNLA. However, Savimbi managed to transform UNITA into an enduring and influential movement.

THE END OF COLONIAL RULE AND INTENSIFICATION OF THE CIVIL WAR (1974-1977)

In 1974, some 320.000 Portuguese lived permanently in Angola. The Portuguese language was the only official language and the Angolan currency escudos officially had the same value as the Portuguese *escudos*. Regime change in Portugal, after the February 1974 coup d'état in Lisbon, increased hope for colonial liberation in Angola. Thus, the three liberation movements intensified their fight against colonial rule but also against each other. The MPLA renamed its armed guerrilla FAPLA (*Forças Armadas Populares de Libertação de Angola*) in 1974 and after Independence in 1975, FAPLA functioned as Angola's official Armed Forces.

Heavily pressured by the Organization for African Unity (OAU, today the African Union) UNITA's Savimbi, the MPLA's Neto and the FNLA's Roberto met in Luanda in January 1975 to negotiate an interim government and organize elections before Independence Day which had been set for 11 November 1975. However, from from February 1975 onwards, heavy fighting broke out between UNITA and both the FNLA and the MPLA's FAPLA troops. This scared the 320.000 Portuguese living in Angola and by Independence, some 300.000 had left. The Polish news correspondent and writer Ryszard Kapuscinski wrote an entire book *Another Day of Life*, describing the mood in Luanda during that summer of 1975: "People ran around nervously, in a hurry, wrapping up thousands of matters. Clear out as quickly as possible, escape in time, before the first wave of deadly air intrudes upon the city". Before their rushed exodus, many Portuguese destroyed materials and infrastructure to make sure Angolans could not benefit from their work and investments. In addition, Angola was left with a lack of educated personnel to govern their own country.

During these crucial months in 1975, the war for Independence transformed into a proxy war between the communist world and the capitalist world. The USSR supported the MPLA with weapons, ammunition and military equipment. Cuba sent 25.000 troops to Angola in 1975-76 to support the MPLA. The USA supplied both the FNLA and UNITA with armaments. This began in 1974 when Secretary of State Henry Kissinger covertly approved 25 million dollars in military assistance. Due to heavy losses of American soldiers in the Vietnam War, the American congress would not send American troops to Angola. Zaïre under Mobuto provided the FNLA with both paratroops and armored cars.

The South African Apartheid regime supported Savimbi and UNITA against the MPLA from the very beginning, deploying their Special Forces to fight side by side with UNITA troops. This was due in part to the history of Namibia. South Africa had ruled Namibia since the end of the First World War when the *League of Nations* handed over the administration of Deutsch-Südwestafrika to South Africa. From 1948 when South Africa formally adopted the apartheid regime, South West Africa (Namibia) was de facto governed as a region within South Africa. This created an international dispute around the legacy of South African apartheid rule in the country. The MPLA had supported the South West Africa People's Organization (SWAPO) and their fight for Independence in Namibia ever since SWAPO was established in 1960. From 1966 until 1989, the South African Defence Force fought against SWAPO, based on the Angolan side of the frontier. The South African regime also feared a communist takeover in Angola. As a result, South Africa actively used the South West African (Namibian) territory when fighting MPLA in Angola.

To support a UNITA takeover at Independence, the South African Defence Force (SADF) entered Angola in October 1975 from its Namibian occupied territory and made rapid progress toward Luanda. As an answer, Castro sent Cuban Special Forces to halt the South African advance. By the end of 1975, some 25.000 Cuban troops were stationed in Angola to back the MPLA government. Not able to prevent the MPLA from taking power at Independence in 1975, the South African Defence Force continued to fight SWAPO and support UNITA until 1989. In numerous military operations between 1966 and 1989, in the so-called South African Border War, the South African Defence Force attacked SWAPO's bases on the Angolan side of the border.

At Independence on 11 November 1975, the MPLA installed Augustinho Neto as president. The leaders of the two other independence movements, Holden Roberto (FNLA) and Jonas Savimbi (UNITA), however, refused to accept the results. The FNLA and UNITA set up a rival government in Huambo, the second largest city in Angola, situated in the highlands some 600 km from Luanda. The Angolan liberation war transformed to an intense civil war, fueled by the intervention of the two cold war superpowers.

Internal rivalry within the MPLA ranks seriously challenged president Neto's power in May 1977 and consequently changed the nature of the MPLA regime. Nito Alves, a member of the MPLA government, killed six senior member of the party and nearly succeeded in ousting president Neto. The MPLA regime bloodily repressed the attempted coup, jailing Alves and killing thousands of his urban poor supporters. The incident made Neto's MPLA more authoritarian and even more brutal; a regime based solely on oppression and fear.

INTENSIFICATION OF THE PROXY WAR (1979-1989)

Augustinho Neto remained president of Angola and the MPLA until his death in 1979. José Eduardo dos Santos succeeded him. Already in 1961, the then nineteen-year-old dos Santos had shown anti-colonial attitudes to such a degree that he was seriously harassed by the colonial regime. Dos Santos left Angola in 1961 and from his diaspora haunt in the Congo, he collaborated actively with the MPLA. With a USSR scholarship, he travelled to Baku (Azerbajan) where he studied petroleum engineering and communications. Returning to Angola in 1970, he obtained high-level positions within the Soviet friendly and anti-colonial MPLA. Before being elected president in Angola by the regime's politburo in September 1979, dos Santos had served both as Minister of Foreign Affairs, Prime Minister and Minister of Planning. Dos Santos is still (April 2017) the president of Angola, a post he has held uninterruptedly since 1979.

The Soviet Union and Cuba intensified their support of the MPLA and dos Santos during the 1980s. Cuban troops increased from 25.000 to 50.000 and the USSR delivered more and better artillery to help the MPLA resist UNITA's attacks. At this time, UNITA and Savimbi were mainly backed by the United States and South Africa.

When Ronald Reagan was inaugurated as president of the United States in January 1981, support of UNITA intensified. From the very start of Reagan's regime, Savimbi was one of his most important African allies against the Soviet Union. Providing weapons and ammunition to UNITA through middlemen and other countries, the United States strengthened UNITA substantially. From 1981 to 1989, Reagan's entire reign, he sustained Savimbi and UNITA with cash, intelligence services from the CIA, military training, and weapons. At the same time, American oil companies operating in Angola paid royalties to the MPLA regime, royalties everyone knew were used to buy arms and pay off Cuban troops for fighting against the US backed UNITA.

In 1988, Reagan invited Savimbi to the White House and his Africa adviser Chester Crocker, described Savimbi as "one of the most talented and charismatic of leaders in modern African history", "a world-class strategic mind" and "a champion of democracy". By that time, it was widely known that UNITA and Savimbi were responsible for the brutal killing of thousands of innocent Angolans.

South Africa also supported UNITA with artillery, training and cash. In addition, South Africa continued to use their own Special Forces to fight the MPLA directly in Angola. In fact, from 1976 to 1988, the South African Defence Force (SADF) carried out twelve major military operations in Angola. South Africa also helped reorganise UNITA's forces, arm UNITA troops, and train its leadership. The most severe battles between the MPLA (supported by Cuban troops and sophisticated armaments from the Soviet Union) and UNITA (supported by the SADF and weapons from the United States) took place mid-1980s. The battle over Cuito (1987-1988), where the MPLA had an important airfield, was considered one of the bloodiest battles in the entire civil war. Another intense battle was fought around the town of Jamba, where UNITA's headquarters were situated from 1976 to 1992. Here the South African Defence Force played a major role in keeping the MPLA out.

THE END OF THE PROXY WAR AND A CHANGE IN THE CIVIL WAR (1989-2002)

After numerous failed ceasefire agreements during the 1980s, a peace accord was signed in New York in late December of 1988, ending the proxy war in Angola. In the name of *perestroika* (restructuring) and *glasnost* (openness) introduced in the USSR by President Gorbachev in 1986 and 1988 respectively, the appeasement of the Cold War started. This influenced seriously the armed conflict between the MPLA's government FAPLA forces and Savimbi's UNITA forces in Angola. One major point in the deal was that Cuban troops should withdraw from Angola and South Africa should exit Namibia. Both actually happened. In January 1989 Cuban troops started to leave Angola, and on 20 March 1990, Namibia was granted Independence from South Africa. The US ended its funding of UNITA and the USSR/Russia stopped supporting the MPLA. Thus, the proxy war in Angola ended with the Cold War and, in a climate of peaceful hope, Angolan society prepared for the first free presidential elections (to be held in 1992).

The 58-page Bicesse accord between the MPLA and UNITA, signed in Lisbon on 31 May 1991, included an agreement to merge the 120.000 troops of FAPLA (MPLA) with the 50.000 UNITA guerrilla troops to form a neutral national army of 50.000 troops. However, the agreement never fully materialised and was completely broken after the first round of presidential elections in September 1992.

Even if the MPLA had declared their Marxist-Leninist ideology dead and opted for a more palatable platform of social democracy prior to the 1992 presidential elections, most Western reporters and researchers, together with Savimbi himself, were convinced that Savimbi would be elected. However, in the first presidential elections ever held in Angola, José Eduardo dos Santos received 49% of the votes, compared to Savimbi's 40%. Savimbi, controlling around 2/3 of Angola at the time of the elections, refused to acknowledge the results. During negotiations for a second round of elections, on the last weekend of October 1992, the MPLA sparked off a three-day bloodbath in Luanda. MPLA government forces killed thousands (counts vary from 10.000 to 40.000) of FNLA and UNITA supporters, including the chief negotiators from UNITA. This resulted in a new and extremely deadly phase of the civil war. While dos Santos and Savimbi signed numerous peace deals during the 1990s, some lasting for hours others for months, not one was successful.

Without foreign support after the end of the Cold War, UNITA financed its operations mainly by selling illicit diamonds. UNITA attacked mines, kidnapped expatriates, disrupted diamond transport routes, and gained control over mines. In the 1990s UNITA "blood diamond"[4] sales reached close to US $ 4 billion, far outstripping the official sale of Angolan diamonds.

On the other hand, the MPLA paid for the civil war mainly with revenues from the oil sector. Some 75-90% of state revenues were generated by oil by the end of the 1990s. The MPLA regime's income ranged from 1,8 billion to 3,0 billion US$/year from the oil sector. Still the Angolan official debt reached 11 billion US$ by 1995. Huge loans were granted to the MPLA based on future oil revenues, and the MPLA government used huge parts of these loans to procure military equipment, mainly from Russia and Brazil.

Not until Angolan Armed Forces killed Savimbi in a battle on 22 February 2002, was a lasting peace deal signed (4 April). Savimbi's obituary in *The Guardian* read that "for the past 10 years, using the proceeds of smuggled diamonds from eastern and central Angola, he [Savimbi] fought an increasingly pointless and personal bush war against the elected government in which hundreds of thousands of peasants were killed, wounded, displaced, or starved to death."

The 27-year-long civil war killed some one million people and around four million had to flee their homes, many hundred thousands seeking refuge in neighbouring countries. In addition, the civil war left Angola with an estimated ten million landmines. While various NGOs are now removing them slowly, landmines have killed thousands and injured at least 80.000 individuals in Angola. Along with the high human cost, close to all

---

4 See e g. UN Security Council Resolutions 1173 and 1176 (1998) for the term "blood diamonds".

bridges in the country were destroyed, four out of five schools were devastated, and a majority of factories, hospitals and roads were destroyed.

THE NEW ANGOLA UNDER MPLA AND DOS SANTOS RULE (2002-)

With Savimbi dead and a peace deal signed, life started to improve in Angola, at least for the elites. Angola's oil production rose steadily and has more than doubled since the end of the civil war, reaching close to 2 million barrels/day as of 2015. In 2008, Angola bypassed Nigeria as Africa's top oil producer and Angola is the single largest exporter of crude oil to China. In fact, around 70% of the Gross National Product derives from oil. Huge investments in infrastructure also helped, and Angola is considered the world's fastest growing economy in the 2000-2010 decade.

However, only a tiny group of elites in Angola took part in this economic miracle, making Angola one of the best empirical examples of the paradox of plenty or the resource curse. By 2004, seven Angolans were worth more than 100 million US dollars. All of them were government employed. One of them, Isabel dos Santos, the eldest daughter of Angola's ruling president, was involved in a conglomerate of various businesses including real estate, banking, telecom and oil, and is Africa's richest woman. Nominated CE of the state-run oil company Sonangol in 2016, she will probably increase her 3,4 billion US$ fortune.

Massive corruption represents a major challenge to development in Angola as do unequal distribution of income and wealth. In 2015 *Transparency International* ranked Angola 163/168 in their worldwide corruption index, while he World Bank ranks it 169th out of 175 countries in their inequality index. In fact, while every year Luanda is ranked among the most expensive capital cities for expatriates worldwide, more than 65% of the population live on less than 3$/day. While the MPLA regime has built some schools and hospitals, as well as roads and bridges, Angola has only slightly improved on the United Nations Development Index since 2002. Infant mortality is among the highest in the world and around one third of the population is still unable to read and write. Hopefully, we will not be witness to another civil war in Angola. Yet, to appease the population the regime needs to supply more than oppression and fear of a new war. The regime needs to distribute economic resources and political power more equally.

REFERENCES

Bissonnette, Brian (1991). *The Angolan Proxy War: a study of foreign intervention and its impact on war fighting*, Ohio: Bowling Green.

Burmingham, David (2015). *A Short History of Modern Angola*, London: Hurst & Company.

Chabal, Patrick (2002). *A Postcolonial History of Lusophone Africa*, London: Hurst & Company.

Crocker, Chester (1993). *High Noon in Southern Africa: Making Peace in a Rough Neighbourhood*, New York: Norton.

Gleijeses, Piero (2013). *Visions of Freedom: Havana, Washington, Pretoria, and the Struggle for Southern Africa, 1976-1991*, Chapel Hill: The University of North Carolina Press.

Gonzalez, Adrian (2010). "Petroleum and its Impact on Three Wars in Africa: Angola, Nigeria and Sudan," in *Journal of Peace, Conflict and Development*, issue 16.

Kapuscinski, Ryszard ([1976] 1987). *Another Day of Life*. London: Penguin Classics

Malaquias, Assis (2007). "Angola: How to lose a guerilla war," pp. 199-220. In M. Bøås & K. Dunn (eds.) *African Guerrillas*, London: Lynne Rienner.

Meijer, Guus (ed.) (2004). *From Military Peace to Social Justice? The Angolan Peace Process*. London.

Meredith, Martin (2006 [2005]). *The State of Africa.* London: FreePress.

Savimbi, Jonas (1986). "The war against Soviet colonialism. The Strategy and Tactics of Anti-Communist Resistance," In *Policy Review* pp. 18-25.

Sean Cleary (1999). "Angola – A case study of private military involvement," in Cilliers, J. & Mason, P. (eds), *Peace, Profit and Plunder: The Privatisation of Security in War-Torn African Societies*. Institute for Strategic Studies: Pretoria.

"They said we have to forgive each other":
Memory, 'Transitional Justice', and (Post)colonialism in the Context of the International Screenings of the Documentary *My Heart of Darkness*

KAYA DE WOLFF

INTRODUCTION

Almost two decades ago, in his seminal anthology *Memory and the Postcolony: African Anthropology and the Critique of Power*, Richard Werbner (1998, 1) declared a "postcolonial memory crisis" emerging widely across the African continent. Werbner argued that this crisis was not "merely about what was publicly remembered or forgotten", but that it concerned rather the "very means and modes of remembrance", whether in everyday life, in major public occasions, or in the context of conflicts and protests (cf. ibid; see also Becker 2011, 321). According to Werbner, the "critique of power in contemporary Africa calls for a theoretically informed anthropology of memory and the making of political subjectivities. The need is to rethink our understanding of the force of memory, its official and unofficial forms, its moves between the personal and the social in postcolonial transformations" (Werbner 1998, 2). In this chapter, I submit that *My Heart of Darkness*, the documentary film which inspired this volume, can be read as an expression of a "popular counter-memory" (Werbner 1998, 1) that gives insights about the modes of official and even more about personal and vernacular forms of memory in the context of post-conflict Angola.

At first view, *My Heart of Darkness* seems to be primarily a documentary film about four veterans of the Angolan civil war (1975-2002) and their struggles to deal with the haunting memories of this traumatic experience. The film centers around the personal experiences of co-director Marius van Niekerk – who also provides the voice-over and who himself was sent to serve in the war in Angola as a South African elite paratrooper – setting off on a journey to come to terms with his past, his memories of the war symbolically stored away in a box of photographs and items from his time in the army.

This box of shameful memories [...] I have to get rid of it once and for all before my daughters discover it. I have to go back and reclaim my life. I have to go back and reclaim my lost innocence. But I can only do this with your help, with your consent. But I am shit scared of what will happen to me, of what you might do to me when I reveal to you my darkest nightmares. Would you be able to forgive me? Would you be able to forgive us for killing you? (3:15-4:17)

This prologue raises the question of who is actually being addressed in *My Heart of Darkness* and who is the collective "us" imagined by Marius. Therefore, the focus of the reflections in this chapter is less on the documentary film itself and more on the controversial discourse surrounding it, as well as the different memories it triggers, all of which is guided by the following set of questions: If *My Heart of Darkness* was supposed to be read as a film about trauma and memory, whose memories are these? If it was a film about reconciliation and forgiveness, (how) does it work? What kind of audience does the film address and what memories does it trigger in the different contexts of its reception at international screenings?

Although the story is dominated by Marius' narration, the documentary has been presented as a "universal film about war veterans" that promotes an optimistic message of hope in terms of reconciliation and forgiveness. Indeed, as this chapter will demonstrate, in the context of screenings in the US and South Africa, *My Heart of Darkness* has been read as a "tale" about war trauma, forgiveness and reconciliation among former enemies and their search for social and psychological rehabilitation, albeit these readings refer to different historical contexts. In the context of screenings in Angola, however, the documentary has been framed as a model of national reconciliation for the post-conflict society. It also presents a glimpse of the different discourses stressing the official national reconciliation scenario.

In light of these divergent contexts, the aim of this case study is to trace the international trajectory of *My Heart of Darkness*, as it has been presented at various festivals around the globe, and to explore the potential of a critical analysis of its local adaptations with regards to power, hegemony and resistance. I argue that both the film and the promotion accompanying its screenings adopt a hegemonic stance in bringing across their central message of forgiveness and reconciliation. Nevertheless, as my empirical study in this paper will demonstrate, the controversial memory discourse within and about the documentary film also provides moments of critique of power and resistance, particularly among an Angolan audience, as it gives glimpse of subaltern narratives. In such a reading, *My Heart of Darkness* speaks to the "postcolonial memory crisis" (Werbner), both to an elite driven top-down-memory politics and a popular counter-memory. Furthermore, it reveals European colonialism's complicity and Western domination of the current Transitional Justice discourse.

As a starting point, I suggest that due to its polysemic structure and the inherent polyphony of the four main protagonists and their narratives, *My Heart of Darkness* becomes an ambivalent cultural text that allows different readings and identifications according to the various contexts and discursive frameworks of its reception at international screenings. Linking cultural media studies and memory studies with postcolonial theory and critique, I will then outline a methodological approach that enables a critical contextualised analysis of *My Heart of Darkness*. More precisely, this means that – following seminal approaches by Stuart Hall and John Fiske – I will approach the film as an open cultural text that evolves its meaning in interaction with an active audience when discourses of the film meet discourses of the audience, emphasizing the issue of power and resistance in these different readings. Furthermore, understanding cinema as a leading media of collective memory allows us to approach *My Heart of Darkness* as a potential "remembrance film" (Erll and Wodianka 2008) and focus on the discursive frameworks and settings that guide its reception and the memories it triggers.

Based on my study of these discursive frameworks, I will present five theoretically informed interpretations of the film's differing contexts of reception, namely in the United States, South Africa, Sweden and Angola itself; all of which, however, refer to readings that are in themselves ambivalent and contested. This approach will demonstrate how *My Heart of Darkness* has the potential to be read as a universal film about war veterans in tradition of anti-war-films; a model for the national reconciliation process; and a platform for emerging critical narratives about the Angolan elite's exploitation of its people. For this is an explorative research design, in my conclusion I will also present the direction further studies might take with regards to the potential adaptations of *My Heart of Darkness* as a "remembrance film" and the memories and debates it might trigger. Finally, I will link my findings back to the discussion of critical media research, memory studies, and post-colonialism in the context of the controversial Transitional Justice process in post-conflict Angola.

In order to provide a context for my argument, in the following, I will briefly draw on current discussions in the field of Transitional Justice about the interplay of memory and reconciliation and outline postcolonial critique of the specific Angolan nation and peace building process.

MEMORY, TRANSITIONAL JUSTICE, AND (POST)COLONIALISM IN THE CONTEXT OF THE ANGOLAN CIVIL WAR

In recent years, the question of memory and reconciliation in the context of post-conflict societies has become a major issue in the interdisciplinary field of Transitional Justice, which has also established a new dialogue between memory studies, postcolonial studies, and genocide studies. In this context, the 'transitional justice discourse', which has become dominant in the context of "transitional" nation and peace building processes, has also raised criticism from postcolonial legal theorists who have pointed to its inherent colonial legacy and neo-colonial aspirations; a perspective that I will evoke as a critical lens on post-conflict Angola.

In its origins "Transitional Justice" designates a repertoire of different tools, strategies, and instruments to help societies pass from conflict to peace in the aftermath of large-scale human rights abuses.[1]

Transitional justice refers to the set of judicial and non-judicial measures that have been implemented by different countries in order to redress the legacies of massive human rights abuses. These measures include criminal prosecutions, truth commissions, reparations programs, and various kinds of institutional reforms. Transitional justice is not a 'special' kind of justice, but an approach to achieving justice in times of transition from conflict and/or state repression. By trying to achieve accountability and redressing victims, transitional justice provides recognition of the rights of victims, promotes civic trust and strengthens the democratic rule of law.[2]

In this view, Transitional Justice (hereafter: TJ) is intrinsically linked to the issue of memory and future-building which has been implied in much of the work done in cultural memory studies in the last years. The TJ discourse makes these issues explicit as it is directly informed "by the belief that future peace and stability depends crucially on finding ways of 'coming to terms' with past violence", as memory scholar Ann Rigney (2012, 251) puts it.[3] With regards to the bourgeoning TJ discourse and the "reconciliation paradigm" (Short 2005, 268), Rigney critically observes the shift towards a new discourse, foregrounding the

---

[1] The term "transitional justice" was first coined by legal scholar Ruti Teitel in 1991, although the roots of the concept can be traced back much further (see Teitel 2008).

[2] Quoted from The International Center for Transitional Justice (ICTJ), see website https://www.ictj.org/about/transitional-justice (last accessed December 8, 2015)

[3] As Rigney (2012, 252) states: "As a huge body of literature demonstrates, post-conflict nation building has become inseparable from the perceived need to come to terms with the divisive legacy of the past so as to generate solidarity or, at the very least, conditions for peaceful co-existence. Indeed, the very idea of 'transition' implies the symbolic inauguration of a liminal period marking the passage from old to new and providing a delimited space for negotiation (Wilson, 2001)".

"performative dimensions" of remembrance and replacing the former paradigm of "trauma" that dominated the memory discourse in the decades following the Second World War.

However, as Rigney also emphasizes, the new 'reconciliation paradigm' recognizes the importance of memory along with its potentially disruptive power; thus "calls for remembrance in the cause of reconciliation are in effect paradoxically calls for its containment" (Rigney 2012, 252; also see Shaw 2007). In this regard, it is noteworthy that – whereas the South African Truth and Reconciliation Commission (TRC)[4] has been famously promoted as a promising legal justice model in transitional contexts –, many other Sub-Saharan African states, such as Angola or Mozambique, have opted for amnesty instead of competing Truth and Reconciliation Commissions and International Courts (see Ameida, Sanches and Raimundo 2010, 3).

A number of recent studies have been dedicated to the 'Angolan case', seeking to interrogate the complexity of the decade-long war and its various struggles, as well as the post-conflict peace-building processes (for an overview see for example Casertano 2013; Koloma Beck 2012; Malca and Skaar 2015). As Teresa Koloma Beck points out in her study *The Normality of Civil War*, Angola looks back upon "one of the most extended war periods in the modern history of the African continent" (Koloma Beck 2012, 21).[5] This period began as a struggle for independence from Portugal

---

4 On the South African Truth and Reconciliation Commission (TRC) see Wilson, 2001; on the TRC and alternative artistic modes of remembrance in post-Apartheid South Africa see for example Buikema, 2012.

5 Since the process of normalization of the Angolan civil-war – meaning the expansion of violence into everyday life – affected both armed groups and the civilian population, Kolomba Beck looks beyond simplistic notions of victims and perpetrators to reveal the complex, shifting interdependencies that emerged during wartime. This blurring of boundaries is also echoed in My Heart Darkness in a slide quoting Philip Zimbardo (who was the author the so-called Stanford-prison-experiment): "The line between good and evil is permeable and almost anyone can be induced to cross it when pressured by situational forces" (00:19).

and developed into a drawn-out decade-long civil war that deeply affecting everyday life of both armed groups and the civilian population.[6]

It was only in April 2002 that the "Luena Memorendum of Understanding" brought a formal end to Angola's conflict, following twenty-seven years of civil war (cf. Ameida, Sanches and Raimundo 2010; Koloma Beck 2012, 21ff; Malca and Skaar 2015). Nevertheless, the present peace-building process has also been criticized by scholars who argue that Angola has, in fact, made no efforts to address the violations committed during the extended period of postcolonial war, in which hundreds of thousands of people died (cf. Malca and Skaar 2015, 149).

In view of the implementation of amnesty laws, Malca and Skaar conclude that the Luena Memorandum of Understanding – like the two earlier peace accords – did not aim to initiate a transitional justice process in Angola but rather helped to foreclose any public debate about the question about accountability:[7] "Its main objective was simply to end the armed conflict between UNITA and MPLA and to facilitate political reconciliation" (Malca and Skaar 2015, 157).

As Skaar and Serra Van-Dúnem (2006, 14) conclude in their earlier study on the role of courts as vehicles for social transformation and justice with respect to the "marginalized and poor" in post-conflict Angola, "the failure to implement social and economic rights in Angola is not primarily due to constitutional limitations, but rather due to the lack of resources among the poor as well as to lack of human and technical resources within

---

6 In fact, the violent conflict began with the anti-colonial struggle in the late 1950s and early 1960s: Angola fought for liberation from 1964 to 1974; after gaining independence from Portugal in 10 November 1975, Angola experienced a "problematic postcolonial transition to peace and democracy", which was only completed in 2002, after two earlier attempts had failed (Ameida, Sanches and Raimundo 2010, 4; Malca and Skaar 2015). In the course of the three peace agreements – the Bicesse Accords (1991), Lusaka Protocol (1994), and the Luena Memorandum of Understanding (2002) – Angola approved a series of amnesty laws that were supposed to lead the national peace building process.

7 In sum, according to Skaar, Malca and Eide (2015, 13), two sets of claims regarding broad amnesties can be distinguished: "that they are good for peace and democracy in the short run, and that they are detrimental to peace and democracy in the long run".

the justice system itself".[8] Moreover, due to the complexity of the Angolan civil war as both a postcolonial civil conflict and an international conflict, there is "an unresolved legal discussion as to which crimes could or should be prosecuted under international humanitarian law" (Malca and Skaar 2015, 167, end note 2).

The argument that Malca and Skaar pursue here is that, even though amnesty does not necessarily mean amnesia, there has not been an institutionalized public discourse providing recognition to the victims of the Angolan wars, their memories, and their social situation in post-conflict Angola. Moreover, the newly enriched country's wealth – coming from the Angola's crude oil industry – is benefitting only a few elites, turning Angola into one of the countries with the largest social inequality among post-colonial states in sub-Sahara Africa, a situation characterized by the term "the two Angolas".[9]

---

8   With regards to the "marginalized" and "poor", Skaar and Van-Dúnem (2006, 8) summarize: "In Angola, the vast majority of the people may be categorized as marginalized in the sense that they are poor and thus lack adequate resources. For this reason alone, it is important to know what kind of resources the poor actually possess and what chance they have of approaching the formal judicial system for recourse when their social and economic rights are violated."

9   "Thus, although the economy is said to be booming, especially the windfall from the oil revenue, little of this trickles down to the masses. Most of it ends up in Luanda and its environs – being shared among the corrupt elite and their families, friends and foreigners. It is estimated that a quarter of the billions in oil revenue goes missing every year without it being accounted for." Alexactus T Kaure "Post-Savimbi Politics in Angola", The Namibian, December 1, 2015, http://www.namibian.com.na/index.php?page=read&id=34677 (last accessed December 11, 2015).

The international community has been criticized as, due to interests in Angola's oil reserves, it has not put any pressure on Angola regarding its gross social inequality, corruption and lack of democratic structures. In fact, China and the USA are presently the two major importers of Angola's crude oil (see Malca and Skaar 2015, endnote 24; also see Casertano 2013). In general, European governments, too, seem to be more concerned with granting stability than democracy in Angola (see Skaar and Van-Dúnem 2006).

In many ways, the Angolan civil war has been characterized as a genuine postcolonial conflict, following the liberation war which was fought by three nationalist movements, each seeking dominance over the newly independent country: Frente Nacional de la Libertação de Angola (FNLA), Movimento Popular de Libertação de Angola (MPLA), and União Nacional para a Independência Total de Angola (UNITA) (cf. Ameida, Sanches and Raimundo 2010, 4). Since independence, "ethnic and political claims have been structured into a system of 'ethnic parties' that influenced Angola's civil life", as Stefano Casertano (2013, 225) points out, who also emphasizes the Angolan conflict as "a perfect case of oil-fuelled conflict" (Casertano 2013, 223) that is characteristic of conflicts in resource-rich (postcolonial) states that are marked by ethnic fragmentation (also see Casertano 2013, 223-231).[10] Koloma Beck (2012, 21) concludes: "Each group had been formed in a particular social and ethno-linguistic milieu, each was supported by different foreign agents", who "fuelled the conflict between the nationalist movements" (Ameida, Sanches and Raimundo 2010, 4), turning Angola into one of the final stages of a Cold War struggle (see Casertano 2013, 220).[11]

As Casertano (2013, 230) summarizes with regards to the post-colonial conflict being fuelled by the oil industry (together with diamonds): "The

---

10 As Casertano (2013, 228) further explains: "In Angola, the main reason for conflict was ethnic: tension had been building during the colonial years, given the different roles and connections of the main ethnicities in the Portuguese-led society. As colonial rule was relieved, conflict exploded – yet oil (together with diamonds) played a significant part in the outbreak of the war".

11 The Soviet Union, together with Cuba and the Eastern Bloc, supported the Marxist-Leninist MPLA with unprecedented military and human resources; regionally, MPLA was backed by Congo (Brazzaville) and Zambia. The FNLA received support from the USA and South Africa, but also from China, the Organisation of African Unity (OAU), and Zaire. The supporters of UNITA, especially the USA and South Africa, had meanwhile tried to defeat the governmental forces of the MPLA through military operations in the south and southeast of Angola (cf. Ameida, Sanches and Raimundo 2010, 4). An equivalent explanation is also provided in the documentary My Heart of Darkness, when Marius outlines the history of the Angolan civil-war in the introduction.

Angolan case is a clear example of the consequences that a country may suffer when the distribution of resource revenue is not aimed at real development, but interrelates with tribal hostility which, in this case, traces back to the colonial rule".

In her insightful essay "Feminism, Postcolonial Legal Theory and Transitional Justice: A Critique of Current Trends" (2012), postcolonial feminist human rights scholar Khanyisela Moyo points out several crucial aspects of the dominant transitional justice discourse. Moyo bases her argument on seminal studies by legal scholars Antony Anghie (and Siba N'Zatioula Grovogui) who, drawing on Edward Said's seminal theory of Orientalism, pointed out "that most international law's core doctrines are a product of efforts by European jurists to legally account for the imperial distinctions between the civilized European world and the non-civilised non-European world" and demonstrated "that European viewpoints of the self and their philosophical depictions have been essential in the formation of international law" (Moyo 2012, 251).[12] Applying this critical lens onto the field of TJ, Moyo argues that such a postcolonial critique reveals the imperial legacies in international law both in colonial as well as (post-colonial) transitional contexts.

Postcolonial critique of the dominant transitional justice discourse exposes it as "one of the human rights strategies that are reminiscent or imperial intervention in the lives of postcolonial subjects"; nevertheless, as Moyo (2012, 263) concludes, it might also include a "liberation potential" since it is "open to seizure by the same [postcolonial subjects, KdW]. This is possible in a transitional context since these situations create opportunities for stakeholders to rethink the inadequacies of the accepted discourse, and to subscribe to new ways of seeking justice".

In this regard, according to political philosopher Nancy Fraser, one of the major challenges and dilemmas of social justice projects is in striking a balance between commitments to redistribution and recognition (see Fraser 1997; Fraser and Honneth 2004). Thus, within the context of transitional justice processes, according to Moyo (2012, 245) "'justice' as recognition

---

12 Furthermore, following Makau Wa Mutua, Moyo (2012, 251) points out that the majority of cases that are being dealt with by human rights bodies are from postcolonial countries: "[I]t is in this sense that the 'other culture', that which is non-European, is the savage in the human rights corpus and discourse".

may involve prosecutions, the creation of truth commissions or commissions of requiry whose role is to recognize or acknowledge the atrocities of the past regime. By contrast, justice as 'redistribution' seeks to re-order property or land rights and can symbolically redistribute shame, from the victim to the perpetrator". Adapting this perspective onto the Angolan context – where arguably none of these actions has taken place so far – the question that comes to the fore here is: who are the victims of the decades-long Angolan civil war that basically involved the whole country? And who are its perpetrators, keeping in mind that the conflict was turned into a Cold War struggle that strongly involved external international actors? Who is, therefore, to be hold responsible, today?

Hence, against the backdrop of a postcolonial legal critique of the TJ discourse, I will approach My Heart of Darkness as a text embedded into current memory discourses related to post-conflict Angola as well as, more generally, in current discourses about memory and reconciliation.

APPROACHING (REMEMBRANCE) FILMS AS OPEN CULTURAL TEXTS AND 'MEMORY CUES'

The methodological approach chosen for this case study builds on work in cultural media studies and memory (media) studies that shifts the focus beyond the film text itself onto the social contexts and discursive frameworks that guide the audiences' readings.

According to Stuart Hall's famous "encoding/decoding-model" (cf. Hall 1980) – which laid the foundation for reception research in British cultural media studies[13] – different preferred readings emerge depending on

---

13 In the tradition of British cultural (media) studies, as elaborated since the 1960s at the Birmingham Centre for Cultural Studies, media research and studies of film and television, in particular, focus on social contexts and subjects. Cultural studies thus departs from a "uses-and-gratification-approach" and instead moves towards an active audience, approaching media products such as film and television as open cultural texts that evolve meaning in a productive interaction when the discourses of the film meet the discourses of the respective audience. From this perspective, film analysis becomes a critical cultural analysis of social relations, power, hegemony and resistance that transcends the film text itself,

the contexts of a film's reception. Hall distinguished between three ideal-typical position from which a media text might be decoded: First, the "dominant-hegemonic-position," which correlates to the dominant ideological system; second, the "negotiated position"; third, the "oppositional position" (cf. Hall 1980, 136). Hall thus speaks of a hegemonic process in the Gramscian sense when the readings of the audience correlate with the suggested messages of the producers and a dominant group wins a subordinated group's voluntary consent to and approval of their 'definition' of social and political events (cf. Winter 2003, 155). However, at the same time, Hall – trained in semiotics and most of all informed by the work of Volosinov[14] – emphasizes that medial messages are always characterized by a polysemic structure, which means that texts might always be interpreted in another way (albeit they are not completely open).[15]

For John Fiske who further elaborated on Hall's encoding-decoding-model, however, *class* is not the central social factor in the analysis of decodings. According to Fiske, identities are not determined by a fixed social position, but rather result from symbolic negotiations in discourses. Therefore, Fiske suggests not focusing on a singular preferred reading (meaning) in media texts but instead to speak of "preference structures" that foster certain meanings and marginalize others (cf. Winter 2003,

---

shifting the focus onto the discursive frameworks of the film's reception in different social contexts and with different audiences.

14 In his seminal work on Marxism and Language Philosophy (Marxismus und Sprachphilosophie, 1975), V.N. Volosinov pointed out that even a single symbol already implies several meanings, it is polysemic; thus already a single symbol can become an arena for the struggle between different social groups over its meaning (cf. Winter 2003, 153).

15 This argument concerning the openness of cultural texts and the undeniable perspectivity of their reception and decoding has become groundbreaking for cultural studies media research (cf. Winter 2003, 155). Following Hall, the ethnographic and discourse analytical research of cultural studies focuses on processes of media production and media adaptation situated between the power of the media (producers) and the power of the audience (cf. Winter 2003, 155). Film analysis, in this view, becomes a critical discourse analysis of its reception that allows insights about dominant circulating discourses, social struggles, and power relations.

155).¹⁶ Moreover, Fiske (1997) identifies characteristics of media texts – such as parody, surplus in signification, contradiction, polyphony and intertextuality –, which are crucial for their polysemic structure and thus open them up to different interpretations and connections, (see Fiske 1987, chapter 6; Winter 1992, 74-86; Winter 2003, 155-156). These characteristics create a potential pool of readings that may vary in relevance depending on the social context of the given audience.

In this perspective, the reception and adaptation of media texts such as film and television becomes a contextualized social practice in which the texts are not pre-existing as objects but are actually produced on the basis of social experience and interaction (see Winter 2003, 156).¹⁷ Therefore, the meaning of a given film text cannot be extracted by an analysis of the film text itself, but only through an analysis that pays attention to the social contexts in which the film is being received and interpreted.¹⁸

---

16 Fiske bases his argument on several, mainly ethnographic studies, such as Morley 1980, 1986 and Hobson 1982.

17 According to Fiske, popular culture then is not a forced, commercial product of the cultural industry, but a product of an active and creative process led by the consumers, the audiences (see Winter 2003, 156; also cf. Fiske 2000, Winter and Mikos 2001).

18 In other words, as Rainer Winter puts it: a given film is being constituted only in the moment of interaction with an audience and, thus, the meanings of a film text cannot be separated from the meanings of an audience. Here, we witness a "convergence of textual and social subjects": the meaning as well as the witnessing subject (audience) emerges in the interaction between the discourses of the film text and the social and material discourses beyond the film (see Winter 2003, 156; cf. also Hartley 1996). It is in this view that the focus of Cultural Studies media research shifted from ideology towards an analysis of the social contexts of media reception and the temporary fixations of meaning (see Winter 2003, 156). Media research in perspective of Cultural Studies then turns into a critical analysis of everyday culture and society. As Rainer Winter (2003, 156) points out, such a cultural studies approach guided by an anti-essentialist critical thinking, might be called a "democratic motivated deconstruction": It is in their interrogation of meaning production that cultural studies demonstrate that society is constantly changing, power relations shift, and that there are always possibilities for a democratic transformation.

As Winter (2003, 156-157) suggests, studies might focus on the contextual experiences of film and might conduct ethnographic analyses (also see Winter 1995); also, secondary texts such as press releases, statements by the producers and directors, or film reviews might also be analyzed, as well as tertiary texts such as fan letters, fanzines, etc.[19] Accordingly, in my study of *My Heart of Darkness*, I will adopt such a film-transcending approach that evolves into a comparative critical analysis of secondary texts, focusing on social contexts and raises questions of power, hegemony, and resistance in these discursive settings at the various countries and festivals where the documentary film was screened.

Furthermore, if we adopt such a cultural studies perspective for memory studies, an analysis of the discourses around the documentary film offers insights of present memory discourses and inherent conflicting narratives. In their volume on film and cultural memory, Astrid Erll and Stephanie Wodianka (2008) promote just such an approach to the phenomenon of "remembrance films" (*Erinnerungsfilme*), one that transcends the film texts (as primary texts) itself.

As Erll and Wodianka (2008, 1) argue, remembrance films represent a recent, powerful and international phenomenon. They emphasize that the attribute "remembrance film", however, should not be understood as a distinct genre but as a socio-cultural phenomenon. Just as any other media of cultural memory, films only turn into remembrance films through their social reception and distribution. It is not the subject of the film itself but the 'things' remembered around the film, meaning the discourses emerging

---

19 According to Fiske (1987, 117ff), three levels of texts can be distinguished related to media products such as (popular) film and television. Next to film and television texts, understood as "first texts", different sorts of texts like press releases, interviews and statements from directors and producers or film reviews qualify as "secondary texts"; moreover fan letters or fanzines etc. can be categorized as "third texts". Fiske's view of these texts is guided by a post-structuralist understanding of culture as a complex intertextual network. Intertextuality means, on the one hand, that texts always relate to other texts and that a given social reality is accessible to a great part only on the basis of the texts circulating in a society; on the other hand, intertextuality also points to cultural texts such as film and television as not being closed but themselves part of an ongoing process of meaning production (see Winter 2003, 156/157).

from the reception of the film, which marks its status as a remembrance film (see Erll and Wodianka 2008, 8).[20] In other words: "A remembrance film does not exist per se but – just as cultural memory – it is the result of communicative, social, cultural and historic-political processes and interactions" (Erll and Wodianka 2008, 5, my translation).[21]

Following these studies of the interplay of film and memory, I will address *My Heart of Darkness* as such a socio-cultural phenomenon; more precisely I understand the film as a "memory cue" (Erll and Wodianka 2008, 5) that potentially triggers memories of the Angolan civil war and beyond.

---

20 Therefore, Erll and Wodianka suggest departing from a classical film-immanent analysis and approaching "remembrance films" in view of their "plurimedial constellations" (plurimediale Konstellationen), instead. In doing so, Erll and Wodianka approach their subjects of study focusing on the memory cultural networks and contexts of the respective remembrance films and their plurimedial functions (Erll and Wodianka 2008, 12). In this regard, films might be asked about their potential for social transformation in the context of post-conflict countries with regards to the "restitution of personhood" as Sharlene Swartz suggests, who in her study of recent human rights films (including My Heart of Darkness) focuses on six areas, namely: "dignity, memory, equality, opportunity, means and citizenship amongst those dishonoured by injustice" (Swartz 2013, 1; also see Swartz and Scott, 2012).

21 In this regard, studies might asked about the film's potential for social transformation in the context of post-conflict countries with regards to the "restitution of personhood" as Sharlene Swartz suggests, who in her study of recent human rights films (including My Heart of Darkness) focuses on six areas, namely: "dignity, memory, equality, opportunity, means and citizenship amongst those dishonoured by injustice" (Swartz 2013, 1; Swartz and Scott, 2012).

HEGEMONY AND COUNTER-MEMORY IN THE DIFFERENT READINGS OF *MY HEART OF DARKNESS*

As outlined in the previous section, linking cultural media studies, memory studies, and postcolonial studies offers a new perspective and the analytical tools to critically approach the film and its surrounding discourse as a social phenomenon resulting from competing narratives and memory politics related to the Angolan civil war and the discursive regime of differing audiences. Thus, the interest of the following case study is not exclusively on the specific Angolan context but on the discursive adaptation of the documentary film in the various contexts of its reception. The focus is on the protagonists and their narratives in regard to the identification(s) they offer for the respective audiences. The empirical analysis is based on a sample of several reviews, interviews, and festival reports published in journals, on the film's website, and festival programmes, as well as in comments on the film's Facebook profile. In total, the empirical material analyzed consists of a collection of nine press releases, reviews and interviews in English, German and Portuguese.[22]

In the following, I will contrast four main readings of the film that go along with its different contexts of reception in the US, South Africa, Angola, Sweden and Germany and speak of divergent memory discourses linked to various memory communities in the respective countries, interpretations of the Angolan civil war, and current memory politics in postwar Angola. This explorative case study will thus demonstrate that, although *My Heart of Darkness* fosters certain "preferred readings" (Hall) or "preference structures" (Fiske), there is no fixed meaning inherent to the film text itself. Due to its polysemic structure, including moments of polyphony, contradiction, and inter-textuality, the film text enables various readings in interaction with different audiences. In all five cases the analysis turns into a critical study that provides insights about power relations, dominant narratives, and subaltern memories in the context of

---

22 See the bibliography of all quoted articles at the end of this chapter. Unfortunately, several links and articles have already been removed from online access. Some more data can be found on some Swedish web-pages; however, due to language barriers, I could not include these articles in my sample for this study.

post-conflict Angola and beyond, ranging from memories of racial injustices during Apartheid in South Africa to the role of the colonizers in the persisting conflict between Bantu and San in present-day Angola.

"I THINK AMERICAN AUDIENCES WILL RELATE TO THE EXPERIENCE OF WAR TRAUMA": READING *MY HEART OF DARKNESS* AS A DEPOLITICIZED "UNIVERSAL TALE ABOUT VETERANS"

The most obvious common characteristic regarding the four protagonists of the film is the diagnosis of war trauma, as Marius, Mario, Patrick, and Samuel all suffer from their experiences and mutually recognize and affirm themselves as veterans of the war, left without anything in its aftermath, and disillusioned and disappointed by the parties they fought for. *My Heart of Darkness* might thus be read, according to the dominant reading as suggested by the producers of the documentary, as a film about forgiveness, recognition, dignity, and hope with regards to the veterans' social rehabilitation and a better future. However, as the following analysis of some reviews and interviews published in the context of screenings in the United States and South Africa will show, this reading is stressed by another discourse, brought to the fore by co-director Marius van Niekerk himself, who repeatedly relates his own White South African identity to the memory of the Apartheid regime and the racial injustices and atrocities committed in its name.

At the start of the film, South African veteran Marius van Niekerk, who was sent to Angola at the age of seventeen and served in the South African Defense Force (SADF) as an elite paratrooper, recalls his experiences of the Angolan war and explains how the gnawing memories of his past deeds caused him to lose everything: his self-respect, his land, his reason, his innocence, and his wife. In the prologue, the documentary starts with Marius' journey back to Angola to ask for forgiveness in order to find personal reconciliation and rehabilitation. "What does a man have to do to regain his self worth after losing it? I have lost my family to exile, my sanity to trauma, my innocence to war" (01:17-1:39). Marius now fears losing his daughters if he won't 'come to terms' with his shameful memories of the war. Over the course of *My Heart of Darkness*, he leads three fellow Angolan veterans on a journey along the river Kwando, towards the former battlefields. Each of the four veterans in the group represents one of the former war parties.

Next to Marius, there is Patrick, who served with the MPLA to defend 'his nation', Samuel who was pressed into service in Jonas Savimbi's UNITA as a child soldier, and Mario, a South African indigene (San) who was recruited by the Portuguese army and then by the South Africans, and whose affiliation was thus constantly changing over the course of the Angolan conflict. Despite their different affiliations and experiences, the common denominator between the four veterans is post-traumatic stress disorder (PSTD), their mutual recognition as veterans, and the acceptance that – despite the atrocities they committed – they, too, were all victims more or less forced into this war, and left on their own in its aftermath.[23]

This reading is also fostered by the narration of the film, as Marius states in the opening scene: "We are all here for the same reason, driven by the need to understand. To set off on a frightening journey into the darkest of our hearts" (5:56-6:14). In many instances, however, the three Angolan veterans seem quite hesitant and the common destiny and unity seems at least partly staged, driven mostly by Marius intervention and his dominant voice-over. This staging, however, opens up moments of contradiction and negotiation concerning the entire project. For example, the three Angolan veterans insist that locating the battlefields where each of them fought is of crucial importance to them; they refuse to accept Marius' argument that he often didn't know where he was being brought.[24]

---

23 It soon turns out that – like Marius – Patrick has also been diagnosed with PTSD. Actually, Patrick has been to the doctor who told him he was suffering from war trauma and should therefore not drink. See My Heart of Darkness, 26:10. Even though Samuel and Mario don't explicitly diagnose their problems as such, and though they seem to have made no attempts to seek dialogue, reconciliation or forgiveness themselves, they, too, seem traumatized and still caught up in the conflict and the aftermath of the war.

24 See the sequence when Marius explains: "I've told you, I think.. It's difficult to try and say where we have always been because the unit I worked in was a special force unit. The parachute regiment. They didn't always tell us where we would go. We were dropped at night, deep into Angola. Sometime we were driven in, nobody told us where we were going, because our operations were

Despite this initial distrust, in a series of river conversations over the course of journey, the four veterans exchange memories of the war and recall the atrocities they witnessed and committed themselves. Through these shared memories, a new unity seems to grow, and the message of joint victimhood and reconciliation becomes more convincing. Eventually, when Patrick states, "What Marius is trying to do... For me, it's very important. For me, it's very important. I know we need that, too" (41:14-41:20), Samuel and Mario also agree.

In this view, a summary of *My Heart of Darkness* promoted by "German documentaries" suggests the following: "In higher political circles, debate still continues on the issue of who won the Angolan civil war (1975-2002), the Cold War's final theater of conflict [...]. But the victims of the conflict know the truth: nobody won. And so do the soldiers who created the victims, as this gripping personal documentary testifies".[25] In this view, *My Heart of Darkness* can be read in the tradition of the anti-war film, and indeed as a rather "universal film about war veterans" that abstracts precise historical and political information about the Angolan civil war. To foster such a reading of *My Heart of Darkness*, the producer's of the documentary, in fact, sought to repress a great deal of the historical and political contexts. As director Staffan Julén said of the documentary on the occasion of the documentaries world premiere at the International Documentary Film Festival Amsterdam (IDFA) in November 2010:

We wanted to make it a universal film about war veterans and what war makes of you... After a while, my hope was that the audience would stop thinking about whom so-and-so was fighting for. Because what mattered was what they did and how they dealt with it. Then we tried to limit the historical information as much as we could, and frame the story through Marius' point of view. To tell what he knows about it, and also what he doesn't know about it.[26]

---

secret". Patrick then intervenes and insists "but you got a map" and only seems partly convinced when Marius negates this "Not us. Maybe the captain, but not the troops. We were just troops" (32:57-34:20).

25 Quoted from German documentaries.
26 Quoted from Filmmaker Magazine.

Moreover, regarding the distribution of the film in the US, the agent in charge asked the directors to provide another voice-over and re-cut the film for its US adaptation, making it more accessible and attractive to a US audience who might not be particularly interested in the specific experiences of the Angolan Civil War. "Their reasoning was that it would be better for the American market if we had a known American actor [doing the voiceover]", explained van Niekerk in an interview with Filmmaker Magazine.

The director's refusal of their US agent's demands to work with an US-American retelling (voiceover) cost them planned screenings at the major festivals Sundance, Hot Docs and, Tribeca. Eventually, the documentary had its US premiere at the Pan African Film Festival (PAFF) in Los Angeles in February 2012, where it was nominated best documentary. The PAFF obviously provides a different context than other Northern American film festivals, as van Niekerk himself emphasizes in an interview. In anticipation of the response of an American audience, he pointed out his presence as one of the few *'white'* directors in competition at the festival, fearing that they would relate him to the Apartheid regime and (falsely) accuse him of being a 'racist South African Afrikaner':

When I looked at the catalogue and saw all the films and people invited, it looks like I'm the only white person that's going to be there. (Laughs) It's like the black film festival in America. And I'm a bit scared, because I'm going to be standing in front of the audience all on my own after the screening, and all of these black Americans are going to go, 'Apartheid, you with South African racist! What are you doing here in America?!', I don't know. I'm sure I'm wrong.

It is particularly interesting to note how this discourse about racism is of major concern to Marius, and it clearly stems from his personal experiences and memory of the South African Apartheid regime. The historical racial injustices and conflict lines that divided the South African society come instantly to the fore in Marius' anticipation of a US audience at the Pan African Film Festival.[27] This demonstrates his discomfort with his own

---

27 It would be of particular interest, to learn about the reactions towards to screening of My Heart of Darkness at the PAFF; unfortunately, I haven't found any further sources in this regard.

ancestry and his supposed involvement in Apartheid politics as a veteran of the national service.

At another point in this interview, being asked if this was his first visit to the United States, Marius again relates his experiences of meeting a Black American in exile in Sweden and, in telling his anecdote, repeats his fear that he will be (falsely) identified as an "Apartheid, racist Afrikaner":

I remember when I arrived in Sweden – I went to exile in 1985 – I met a black American who was also in exile in Stockholm. And he actually did national service. It was just after Vietnam. When I met black South Africans in Sweden, they were so critical when they saw me as a white, South African, they thought I was an Apartheid, racist Afrikaner. But obviously when they got to know me, and the reason I left, they could relate to my story. But this black American thought my story was amazing. We got along very well. 'Great, Marius! You fought in the war and you turned your back against it!' I'm thinking of that experience I had with this black American veteran in Sweden. I think American veterans are going to relate to this story very much. And hopefully, they will relate to the anti-war side of it. Because a lot of Americans go to war today. It think it's amazing that boys still go and fight. I would think that seeing the causes of the war, especially in America – veterans coming back fucked up and traumatized – there would be a stronger movement against the military in America. But there hasn't been.

The discursive shift in this anecdote is remarkable, as it exemplifies how the interview opens up a discourse about racism, 'black' and 'white' subjects and historical conflict lines both in a US context and with regards to South African Apartheid; however, it actually ends up bringing the anti-war discourse and war veteran's destiny to the fore. In concluding, when Marius is asked about the film's function not only for veterans but for a more general audience of non-veterans, he himself interprets the film's central message in terms of anti-war-films, reconciliation, and the search for personal forgiveness and rehabilitation. This time he completely avoids the issue of racism and relates his experiences to those of US-veterans of the Vietnam war and their efforts to be forgiven and socially (re-) integrated.

Based on this interview and the discursive turns it takes, it seems that in the US, the film was received in the way the interviewer suggests. US veterans are likely to read the documentary in the tradition of anti-war films and, first and foremost, as a personalized emotional portray of PTSD.28 In this way, *My Heart of Darkness* might be read as a film about the destiny of war veterans around the globe, promoting unity, solidarity and mutual recognition as victims of war. The film pertains to all veterans, be they US combatants fighting in Vietnam, South African ANC resistance fighters, or veterans of the newer wars in Iraq or Afghanistan. Just as the promotional text of the German producers Beetz Film reflects: The reasons for the wars may be diverse but the consequences for the survivors don't differ, no matter if they fought in Vietnam, Iraq or Afghanistan.[29]

Yet, such a hegemonic reading, however, removes important political information, historical contexts, and questions about the war's logic, victimhood and accountability from the discussion. This reading does not provoke further debate, as the audiences simply agree on the war's senselessness. Rather than political and historical contexts, "hope", "dignity" and "forgiveness" become key terms with regards to the social rehabilitation of the veterans and their future lives. Overall, *My Heart of Darkness* is promoted as sending an optimistic message, captured, for example, in this plot summary provided by the German producers: "The veterans return as different persons from this journey. Together the four gained back what they had lost on the battlefield – their humanity and dignity".[30]

---

28 The link and inter-textual reference to Coppola's Apocalypse Now (opening scene) seems crucial in this context, even though it is not explicitly evoked in this interview.

29 "Die Gründe für die Kriege sind unterschiedlich, die Folgen für die Überlebenden unterscheiden sich nicht, egal ob sie im [sic] Vietnam, Irak oder Afghanistan gekämpft haben", quoted from Gebrüder-Beetz film website, http://www.gebrueder-beetz.de/produktionen/mein-herz-der-finsternis. accessed December 1, 2015.

30 My translation, quoted from the Beetz film website: "Die Veteranen kommen als andere Menschen von dieser Reise zurück. Gemeinsam haben die vier das zurück erlangt, was sie auf dem Schlachtfeld verloren haben – ihre Menschlichkeit und Würde". See http://www.gebrueder-beetz.de/produktionen/mein-herz-der-finsternis. accessed December 1, 2015.

"...SOME YEARS AGO WHITE SOLDIERS WERE HERE IN THE TOWNSHIPS KILLING BLACK PEOPLE": READING *MY HEART OF DARKNESS* IN THE CONTEXT OF POST-APARTHEID MEMORY IN SOUTH AFRICA

Contrary to the US context, in South Africa, *My Heart of Darkness* was more likely to be read against the background of the national past with regards to the large-scale racist injustices during Apartheid.[31] In the interview with Filmmaker Magazine quoted above, Marius anticipated that his presence as a 'white' South African who used to serve in the national defense force would trigger memories of Apartheid-era military invasions of 'Black' townships. These invasions also form a crucial part of Marius' own background, which he emphasizes in the documentary when he states: "For me this journey started 28 years ago. That day I said 'no' to the killing of my black brothers and sisters. That day I was forced to leave my homeland South Africa into exile in the cold and peaceful Sweden" (4:22-4:48). The audience learns that Marius was sent to an Afrikaans school in town at the age of six where racist indoctrination started. "We were told that the Blacks were communists and wanted to take our land, to kill our parents and rape our women. So getting into the army was a natural thing to do. It was expected of you, it was our destiny. It was something to defend, something to fight for. But was it worth killing for?" (18:30-20:10).

In the South African context, *My Heart of Darkness* thus shifts and brings up memories of Apartheid and the liberation struggle that are arguably particularly present among a local audience in Soweto.[32] Marius' search for forgiveness might indeed be perceived as an unprecedented project, since, even though South Africa has installed the TRC, there has never been a broad public dialogue between the different communities, or even less a collective apology issued from the formerly privileged part of society to the formerly unprivileged for past racial injustices (and continuing injustices in the present). Hence, even though Marius deserted from military service, an action that meant life in exile, he fears subjection

---

31 On the history of Apartheid (1948-1994) in South Africa, see for example Dubow, 2014.

32 Among others, Soweto occupies a particular role in collective memory of Apartheid with regards to the Soweto riots in 1976.

as an ex-military who served in the national suppression of anti-Apartheid resistance.

> The first screening [of the Tri Continental Human Rights Film Festival] was in Soweto outside Johannesburg. During apartheid, Nelson Mandela and all of these activists of the African National Congress (ANC) were living there. The heart of the liberation struggle was born in Soweto. During my time in national service, the white troops were sent into the black townships to fight the ANC. So in Soweto, I was shit scared because I thought, 'My God, some years ago white soldiers were here in the townships killing black people.' Most of the audience was black. Afterwards it was dead silent. And after ten minutes of standing at the front looking at this audience, one black guy got up – he was about my age, probably in the war on the other side – and he was crying, and gave me a hug in front of everybody. And [the audience] was all ANC veterans. They came and hugged me and shook my hand and we had this amazing discussion about the war and reconciliation afterwards. Just incredible.

Contrarily to his fear, according to Marius, the screening of his documentary in Soweto turned into a 'tale' of reconciliation with ANC-veterans and recognition of Marius' individual resistance against the South African Apartheid regime and his personal sacrifice. This kind of restoration of humanity and dignity, or "personhood" (Swartz) seems to be at the heart of Marius' search for "forgiveness" and his mission of "personal reconciliation"; in this view, he thus achieved the central objective of his journey not only back to Angola but also to his native country South Africa, which he had to leave when he refused to further serve the racist Apartheid regime. [33] As Marius explains in the interview for *Filmmaker Magazine*, both in South Africa and Angola he felt he had been accepted and forgiven by local audiences who, in this post-conflict setting, represented the victims of the war.

---

33 With regards to the (re)negotiations of white South African identities, Sally Mathews (2012) emphasizes: "South Africans today live not only with the memory of the racial injustices of the past, but also with present injustices that are a consequence of that past. How should white South Africans live with these past and present injustices?"

After South Africa, I went to Angola and had a couple of screenings at the Luanda International Film Festival. And we got the prize for Best Documentary. That was kind of the end of that road for me, the end of my long journey towards my personal reconciliation. It was an acceptance. I've asked for forgiveness and have been given that. I've been forgiven. So the screenings in South Africa and Angola was the end of that journey for me, in a way. Now, when I have screenings I'm moving on in a different way.

"THEY SAID WE HAVE TO FORGIVE EACH OTHER": *MY HEART OF DARKNESS* AS A MODEL FOR AND/OR AS CRITIQUE OF THE NATIONAL RECONCILIATION PROJECT IN POST-CONFLICT ANGOLA

Contrarily to the initial intention of the producers, *My Heart of Darkness* makes a significant political shift in context in Angola itself where several screenings of the film were promoted under the programmatic title "Reconciliation: Use forgiveness and love your enemy".[34] This promotion framed the journey of the four veterans as a model for the national reconciliation project in Angola as outlined above. However, I argue that whereas the screenings of *My Heart of Darkness* have been used as a popular promotion tool of the national reconciliation project, according to the promises of the Luena Memorendum of Understanding from 2002 (as discussed above), it might also be read as a critique of the same and actually opens up a discourse about the deficiencies of the peace-building process that has failed to open public discourse and to officially recognize the victims of the decades of war and conflict. Indeed, such a critical narrative is inherent in the film text itself, as expressed by the Angolan protagonists with regards to the national memory politics that draw on a discourse of reconciliation and forgiveness. In a sequence at the end of the documentary, Patrick states: "They said we have to forgive each other. Yes, we forgive each other. But I need to know what we were fighting for. We just fight and the war stopped. And we must forgive each other. But I don't know why we were fighting. We were supposed do that before, not now.

---

34 Which reads originally in Portuguese: "Reconciliação: Ousar perdoar e amar seu enimigo".

Why didn't we do that before?" (51:58-52:22). Even though he does not further outline on whom exactly he addresses by "they", he surely refers to the dominant discourse of national reconciliation in post-conflict Angola.

Against the background of the lack of a public platform for the memories of the war, the crucial issue of the screenings of the documentary in Angola seems to be that *My Heart of Darkness* makes the experiences of the war explicit, portraying perpetratorship at all sides of the former war parties, albeit the majority of the veterans were likely to have been recruited by force themselves.

In this regard Marius van Niekerk emphasizes the different contexts of the screening of My Heart of Darkness in South Africa and Angola, linking them to the respective memory politics and particular TJ processes:

The difference between South Africa and Angola is that South Africa had a reconciliation, the Truth and Reconciliation Commission. There was a process there where former soldiers could meet their [victims]. But in Angola, that process has not even started. The two armies in Angola, the MPLA and UNITA, are integrated, but no reconciliation has happened. There are no talks or interest in reconciliation. They don't want to look back. As I was screening the film and had discussions, [the audience] realized the importance of reconciliation and that the different sides of the war have to have a platform to meet and talk. It created a lot of discussion. (Filmmaker Magazine)

In fact, the screenings of the documentary film – in Angola and abroad – have been promoted by the Angolan embassy and supported by national civil service institutions such as the *Ministério dos Antigos Combatentes e Veteranos da Pátria* and the *Instituto Nacional da Criança, sobas, militaries*.[35] As the Angolan Embassy in Sweden announced, *My Heart of Darkness* was screened in as in Stockholm on 4 April 2011, on the

---

35 The ministry for the social integration of the former combatants and veterans, Ministério dos antigos Combatentes e Veteranos da Pátria (MACVP), which seeks to (re)integrate war veterans into society, family life and the labour market, was founded after Angola gained independence from Portugal in 1976. See website: http://www.macvp.gov.ao/Institucionais/Missao.aspx (accessed December 8, 2015).

occasion of the Angolan commemoration day for Peace and National Reconciliation.[36]

In this context, co-producer Tony Nguxi – who is also a popular Angolan musician, singer and songwriter and founder of the philanthropic project "Akwafrica – Africans to Africans for the World" – became a sort of delegate, promoting the documentary abroad and within Angola and strongly emphasized the film's mission of bringing reconciliation to Angola.[37] Nguxi is being quoted in most of the Angolan reviews related to *My Heart of Darkness* and seems to have taken over the role of a foreign correspondent (or ambassador) who connects with Angolan diasporas through his project "Akwafrica" and is thus also entitled to deliver reports about the screening of the documentary in Sweden. To give an example, in an Angolan review, he interprets *My Heart of Darkness* stating that the documentary seeks to inspire everyone in Angolan society to participate in the psycho-mental reconstruction of the children of the veterans and ex-military men.[38] With regards to the screening of the documentary in Sweden in April 2011, Nguxi reported an over-all positive impression of the Angolan Diaspora, stating that many people can relate to the film which portrays the war and sends a central message of forgiveness, peace and reconciliation.[39]

---

36 See Boletim de Informacao n°47, issued by the Angolan Embassy in Sweden.

37 See Tony Nguxi's website http://www.tonynguxi.com "AKWAFRICA is a socio cultural movement of musicians, artists, journalists and entrepreneurs who are determined to give part of their time, energy and professional dynamics in order to create a clear inter-linkage between immeasurable spiritual values that carry all human being throughout the world to be channeled on the socio-cultural and ecologic interventionism in the Cradle of mankind for the benefit of Planet Earth. This project was established in the mid 90s, of the XX century, and founded by the Angolan musician, Tony Nguxi" (accessed December 8, 2015).

38 See Jornal de Angola, April 19, 2011: "Segundo o co-produtor do filme, o angolano Tony Nguxi, My Heart of Darkness visa inspirar a sociedade angolana para participação de todos a reconstrução psico-mental dos filhos dos veteranos de guerra e dos ex-militares".

39 "Instado a comentar sua percepção sobe os debates em torno do filme, afirmou que deu para medir as reacções. 'Muitas pessoas ainda se revêem neste filme

Furthermore, Nguxi is quoted in a statement that promotes the reconciliation paradigm: "The wars and their disastrous consequences still continue in the memory of the people. It is necessary to opt for new directions. The people search for reconciliation and spiritual healing of the traumas caused by whatever war in whatever part of the world".[40] This vision also points to a new direction in Angolan memory politics, which had been dominated by amnesty laws and thus failed to enter into a public discourse about the war and its consequences. As Marca and Skaar (2015, 149) outline with reference to a "right to truth": "the amnesty law of 2002 does not qualify as a transitional justice measure, because it is incompatible with emerging internationally recognized 'right to truth' and with the long-standing right to justice to which relatives and survivors are entitled". For the civilian society in present-day Angola, the decade-long conflicts continue among the ethnically divided communities. This traumatic afterlife of the wars becomes most obvious in the figure of Samuel, who served the UNITA troops and still fears revenge from former enemies and victims. This hinders any search for forgiveness and reconciliation in terms of a real dialogue. Whereas Samuel first argues that they have already left the past behind, he gives voice to his sorrows when he reasons: "Even today, if I'd drive from here to Cuito and tell the people, 'Sorry, I killed your children…', what would they do to me?" (40:05-40:20).[41]

---

que retrata a Guerra, mas que trás como mensagem principal o perdão, a paz e a reconciliação' (….), afirmou. 'As guerras e as suas consequencias nefastas ainda continuam na memória das pessoas'. É preciso optar por outras vias. As pessoas buscam a reconciliação e a cura spiritual dos traumas causados por qualquer Guerra e em qualquer parte do mundo', sublinhou". (quoted from Platina Line, April 16, 2011).

40 Quoted from Platina Line, April 16, 2011, my translation.

41 In his review, Daniel Lindvall directly links Samuel's insecureness back to the marginalized position of former UNITA soldiers in present-day Angola: "Finally, there is Samuel, who was pressed into Jonas Savimbi's UNITA as a child solider. Savimbi, often described as a psychopath, became a Portuguese agent in the early seventies. Though UNITA was set up as an alternative liberation movement, partially along ethnic lines, they quickly started fighting the MPLA rather than the Portuguese. After independence, as the MPLA looked set to take power, UNITA was financed and armed by South Africa, Israel,

Eventually Samuel dares to share his traumatic memories of the brutal atrocities he witnessed as part of Savimbi's troops, namely, in Jamba he witnessed thousands of women burned alive because they were accused of having bewitched the UNITA soldiers (42.10-44:57).

Overall, the screenings of *My Heart of Darkness* seem to stress the official discourse of national reconciliation, which opted for amnesty and political integration of the MPLA and UNITA but foreclosed any public debate and reconciliation process among the Angolan people. In view of this reasoning, one might also suggest a reading of *My Heart of Darkness* as a critique of the elite driven town-down-memory-politics in post-war Angola. Such a "critique of power" (Werbner) is articulated by the struggling Angolan veterans, mostly Patrick and Samuel, who are still trying to make sense of the war, their specific role in it, and their present situation and positioning in the post-conflict Angolan society; in other words they feel deprived of their "right to truth" (Marca and Skaar).

My *Heart of Darkness* opens up a space for the emergence of a vernacular counter-memory narrative about the Angolan war that critically opposes the official memory discourse of national reconciliation. This critique can thus be read as an accusation against the Angolan elite's exploitation of war servants, and their continued reluctance to admit guilt and responsibility in an appropriate way.

"I NEVER SAW THE SON OF ANY CHIEF FIGHTING": READING *MY HEART OF DARKNESS* AS A MODEL OF 'RECONCILIATION-FROM-BELOW' BASED ON 'CLASS-SOLIDARITY'

Towards the end of the documentary, Patrick, again, points out the fact that no politicians' or military men's sons were involved in the war. Thus, he

---

Zaire's Mobuto and according to the logics of the cold war in those days – much like the Khmer Rouge – by both China and the US. The decades-long war did what it was supposed to, preventing any real chance of creating a 'good example' of leftist rule in the resource-rich country. Like Mario, Samuel appears shy and self-effacing to begin with. Like many former UNITA soldiers he fears revenge."

raises the issue of class privileges and social inequality in present-day Angola as a consequence of the war and the elite's exploitation of the Angolan people. "Wherever I fought, I never saw the son of any chief fighting. It's just the sons of the poor. I'm saying it again. I was a soldier. But while we were in the war, the others could study. That's why so many veterans are thieves nowadays – or crazy" (47:12-47:42). Referring to Patrick's comment, Marius van Niekerk further explains the marginalization of veterans in present-day Angola:

In Angola, all the soldiers that fought have nothing, they're completely fucked up. They're completely traumatized. They're marginalized. And the politicians and the military leaders are sitting with all the money. They've made money out of this war. And that's the type of message that Patrick says in the film: that there's no son of any politician or leader that was fighting in the war. That's my personal message anyway. (Film Maker Magazine)

This line of reasoning brought to the fore by Patrick and Marius was taken up by Swedish reviewer Daniel Lindvall in *Film International* who reads *My Heart of Darkness* first and foremost as a model of "reconciliation-from-below" based on "class solidarity".

As Lindvall observes, whereas at the beginning of their conversations about the Angolan civil war the (Angolan) veterans still discuss their particular affiliations and the logic of the war, partly defending the side they were fighting for – as most obviously in the case of Patrick – they come to admit that they were all forced to serve in the war, and were left disappointed, disillusioned and disorientated in its aftermath.

The men come to realize that no matter what side they fought on, they have much in common. They were young men, boys even, with few opportunities, more or less pressured into participation in this war. Certainly, it is natural that van Niekerk has the greatest burden of guilt to carry, as he had options unavailable to the others.

Lindvall thus juxtaposes the veteran's biographies and affiliations against the background of present discursive settings and power relations in regards to the legitimacy accorded to the former war parties in present-day Angolan (memory) politics. This becomes most obvious in the portrayal of Patrick and his narrative about the war, initially reproducing a nationalist discourse in favor of MPLA politics, which speaks of the present discursive setting in post-conflict Angola where MPLA's involvement in the war is arguably being backed up with the broadest legitimacy.

There is Patrick, who fought with the MPLA to defend 'his nation'. A big bulk of a man, he initially exudes national pride and machismo. Having been on the 'right'/right side of the conflict he takes a rather high moral ground at first and also, loyally, maintains that his life in the army did him nothing but good (I hesitate about quotation marks here – 'right'/right – since the MPLA certainly was and is not without serious faults, still it did historically represent the side I would, though critically, defend). (Film International)

The shift that Patrick's narration takes over the course of the film, then, offers a very different, critical memory related to the experiences of MPLA veterans that is probably given little space in the discursive setting in present-day Angola. When being introduced, Patrick states, "I just went to the war. I was too small, but I went. I don't have anything bad to say about the war. The things I know... I was in the war, I fought, I went home... Finished! No one was bad to me. Everyone was good to me. I loved it" (10:46-11:27). His provocative statements are, however, being contradicted by a slide which states "Patrick Johannes. Veteran from FAPLA MPLA. At 14 recruited by force". A couple of minutes later, the audience actually learns about Patrick's forced recruitment as a child soldier when he recalls: "The soldiers captured me. They asked me to go with them, and I just had to say yes. I was on my way to school...They took us to a training camp in Malange, and then to Cuito Cuanavale. I fought in the battles of Memonge and Mavinga. The first battle was terrible [...]" (16:12-17:01).

Besides providing insights into his war trauma – which had cost him his first wife who had left him due to his acts of domestic violence – on several occasions Patrick brings up war veterans' missed opportunities for schooling (17:48-18:32).

Nevertheless, Patrick keeps up his nationalist loyalties and legitimizes the MPLA's war both vis-à-vis UNITA troops and South African invasion. It becomes obvious that since his early childhood Patrick's environment had been dominated by a narrative of the brutality of UNITA soldiers as well as foreign invasion into Angola, namely by South African and Israeli military, forcing MPLA to defend its own community.[42] Patrick accuses these two countries of invading and recognizes their position as some of the major powers in the Cold-War context.

> But I saw what the enemy did, and that's why I went to fight. I'd seen as child how UNITA blew up bridges, blew up roads, killed people, stabbed pregnant women and killed children. But you South Africans invaded Angola with the Israelis, with superior power. You invaded and bombed with aircraft. Should I have waited for you to take our land? No, I couldn't sit still. I was forced to fight against you. (17:01-17:39)

This narration about the conflict lines between MPLA and UNITA during the post-colonial conflict also speaks to the dominant discourses and lack of interaction across the ethnically divided post-conflict society in present-day Angola. This is, however, increasingly challenged as Patrick learns about Samuel[43] who was also forcefully recruited as a child soldier and thus expresses compassion for his former enemy: "I don't feel as if I'm with the enemy. I think he's like my brother. Because he's innocent. He has no guilt. He was forcefully recruited, like he told us already. So I think... I have a lot of compassion for him" (50:29-50:50).

---

42 For example, Patrick argues with Marius: "You need to apologize. It's your right. Because you came to Angola to fight. And me, I was defending myself" (39:25). It is interesting to note that the Cold-War context of the Angolan conflict is rather absent from the memory of the veterans, with the exception of the role of South Africa and Israel. Patrick accuses these two countries of invading and recognizes their position as some of the major powers in the Cold-War context.

43 Samuel is being introduced by a slide that titles: "SAMUEL MACHADO AMARU. At 15 forcefully recruited by UNITA" (12:41); in My Heart of Darkness Samuel gives an account of the harsh conditions of UNITA fighters over the years of war: "When you see hundreds of people die around you every day, you die inside" (13:04). "All those years in the bush...Our lives wasted away. You had nothing – no shoes, no salt, nothing. You eat sand, you sleep badly on grass... That type of life, no... When we could have had a nice life, like other people" (13:12-13:35).

Even though Samuel remains more silent or at least hesitant when being asked if he doesn't feel the need to ask for forgiveness, he eventually also joins the process of reconciliation with his former enemy Patrick and redirects the responsibility towards the government: "I don't think anything, because we were in the war but we're not the ones to be blamed. The government is to blame. They forced us to the front. But now that we're here together, I feel like you're my brother". The sequence finds its 'happy end' with Patrick saying: "Give me your hand, because we're brothers. We didn't deserve war. Let's rebuild our lives and our countries. Be done with the war. Stop the war. It's no good. What happened happened" (50:58-51:50). Even though Patrick's statement partly echoes the national reconciliation discourse, this sequence rather demonstrates the emergence of a new solidarity beyond ethnic divides encouraged through former war parties, and formed in primary opposition to political and military elites.

In this perspective – which might represent that of the Angolan people – the official memory discourse of national reconciliation and current transitional justice process in post-conflict Angola appears as a way of silencing the vernacular memory of the 'Angolan people' and, most of all, the Angolan veterans, who were pressured into this war and did not receive any state support in its aftermath, but were left only with a feeling of personal guilt and shame expressed in their sufferings from PSTD.

Thus, at issue here is neither the precise political context and the veteran's particular affiliations to one of the war parties, nor is it an abstract reading of a "universal film about war veterans"; rather, it is a critique of an elite-driven memory discourse reluctant to address the issue of 'class' and current social injustices. Indeed, one might argue that in view of the amnesty law and impunity, the current course of memory politics foregrounds a sort of "fore-closing of memory", promoted under the headline of "national reconciliation". Instead of opening up a discourse about the experiences of the 'reconciliation paradigm' in the Angolan context, it closes off any remembrance of the mass atrocities committed in the country. Furthermore, the portrait of the Angolan war veterans speaks to the Angolan officials' unwillingness to recognize the victims of the war and redistribute its land and wealth in its aftermath becomes subject of critique.

The process of mutual recognition of their common destinies, portrayed in the documentary, might indeed form a model for "class solidarity" (Lindvall) among the four veterans in *My Heart of Darkness* and even inspire a wider more inclusive solidarity movement among a post-conflict Angolan society that might eventually challenge the official discourse of staged national reconciliation and demand state support in the aftermath of the wars. This emancipative power surely depends on the audiences the documentary reaches, the specific environment guiding its reception, but it could possibly provide a platform for the emergence of a critical narrative shifting the TJ discourse in another direction. This critique of power articulated within the framework of an emerging popular counter-memory might thus eventually result in an argument for the redistributing the benefits of the war among the population in accordance with Fraser's bivalent conception of social justice, as Moyo (2012, 245) suggest when she outlines that the mode of "justice as 'redistribution'" demands to re-order property or land rights; a process that can confront great social inequality in the aftermath of war and conflict in material terms, and furthermore, socially rehabilitate civilian victims (even though they might be former combatants themselves).

"THE WHITES USED US, THREW US AWAY AND TODAY NOBODY CARE ABOUT US": GIVING VOICE TO OUTCAST SAN SOLDIERS AND RAISING THE QUESTION OF THE COLONIZER'S ACCOUNTABILITY

Finally, *My Heart of Darkness* also gives voice to the experiences of the former San soldiers as represented by Mario Mahonga who was recruited by the Portuguese at the age of fifteen and later by the South African army.[44] Even though Mario remains the most quiet of the four veterans and seems rather excluded from the emerging unity between Patrick and Samuel, the documentary does provide a platform for his narrative, linking the Angolan conflict back to its origins in the anti-colonial war period and raising the question of the former colonizers' accountability. Mario

---

44 Slide: "MARIO MAHONGA. At 15 recruited by the Portuguese later by South Africa" (14:47).

explains the situation of his people, the San, who remain marginalized and face discrimination by the Bantu people as a consequence of their service as soldiers for the Portuguese and South African military[45]:

> The Portuguese, Namibian and the South African army all exploited us. They used the bad relation between the Bushmen and the Bantu. The Bantu enslaved us to work for them. My people, the San are a minority group. When the war broke out, the Portuguese convinced us to fight for them to take and rule the land. [...] The Portuguese came to us with false pretensions and misused us for their own objectives and that created more hate between us and the Bantu people and our way of life causing us to loose our self worth. The Bantu people don't see us as equals but think less of us and made us feel regret for being soldiers. The whites used us, threw us away and today nobody care about us. (14:47-15:58)

The accusation that Mario makes explicit here points to the colonial 'divide and rule' strategy that, as in many other contexts of post-colonial states in Africa today, is being identified as one of the causes for ethnical segregation and ongoing conflicts. Lindvall also emphasizes this in his review of *My Heart of Darkness*: "Mario, a Bushman, was recruited first into the Portuguese army, then by the South Africans. The colonial masters, as everywhere, exploited, aggravated or created ethnic tensions and racism in order to divide and rule. The legacy lives on in the form of discrimination. Mario remains the most quiet and solitary of the four men".[46] Although there is little more information on the San in present-day Angola presented in *My Heart of Darkness*, the inclusion of Mario in this portrait of reconciliation opens up a discourse about the colonial legacies within the continuous conflict between the Bantu and the San, the latter being marginalized in the political landscape of present-day Angola as well as expelled to live in remote settlements in South Africa. In the interview with *Filmmaker Magazine*, Marius describes the screening of the film among Mario's community and sketches the present isolated situation of the San soldiers.

---

45 On the particular history and militarization of the San in context of the wars in Angola, Namibia and South African, see Hohmann, 2003.
46 Daniel Lindvall, Film International.

[…] when I was in Johannesburg for the Tri Contintental Human Rights Festival, I went to see Mario who lives in a Bushmen's settlement in Bloemfontein, in the middle of South African countryside. Mario is like the spokesman for the Bushmen, who were dumped out in the sticks by the South African Defense Force after the war in 1994. "There were 5-6,000 veterans and their families in their community, and I had a special screening at their cultural center. And it was just so amazing. All of these people heard that Mario was working on the film, that he went to Angola, but when they actually saw the film it was such an incredible experience. These outcast Bushmen, San soldiers, saw this whole reconciliation process.

My Heart of Darkness thus brings the memory of the war and the militarization of former San soldier to the fore as well as the marginalization that they presently face as "outcast Bushmen" due to the tense San-Bantu-relations resulting from the colonial past. Even though the role of the former colonial powers is not being addressed in the reviews collected for this case study, the screening of the documentary film arguably offers a platform to renegotiate the role of the San soldiers in the Angolan conflict and eventually think of models of reconciliation with the Bantu population.

CONCLUSION AND OUTLOOK

The focus of this chapter has been on the different audiences and readings of the documentary film *My Heart of Darkness* in the context of its international screenings with regards to issues of memory, reconciliation and (post)colonialism. Moving away from a 'classical' film analysis, I have set off to develop a theoretically and political informed methodological approach in the tradition of Cultural Studies critical media research that allows the analysis of the reception and discursive adaptation of the film in its different contexts to turn into an interrogation of circulating hegemonic and subaltern memory discourses, social relations, power, and resistance. When we approach the documentary film as a cultural text that is never "closed" but which unfolds its meaning in interaction with an active audience then an analysis of the documentary and its reception allows us to explore *My Heart of Darkness* as a socio-cultural phenomenon, a "remembrance film" (Erll and Wodianka 2008) triggering diverse memories and contested narratives that are dependent on the dominant discursive frameworks in which it unfolds.

Due to the polysemic structure of the documentary film, *My Heart of Darkness* reveals itself as an ambivalent cultural text that allows very different readings according to the different contexts of its reception in the United States, South Africa, Sweden, Germany, and Angola itself: Firstly, when *My Heart of Darkness* was presented as "a universal film about veterans" that portrayed the veterans suffering from post-traumatic-stress-disorder, it was most likely to trigger memories of the Vietnam-war among a US-audience; secondly, due to Marius' own background as native ('white') South African and an ex-military of the South African army, it also brought up the memory of the brutal Apartheid regime when screened at the Pan-African Film Festival in Los Angeles and even more so in Soweto, South Africa; Thirdly, the documentary film has been framed as a model for national reconciliation towards a domestic audience in Angola and in the Diaspora. While this three readings present "preferred readings" (Hall) according to the producers of the film, *My Heart of Darkness* also enabled more critical negotiated and even oppositional readings with regards to the "transitional" nation and peace building process in post-conflict Angola. As, forthly, the film itself also critiques town-down-memory-politics, accusing the elites' exploitation of their own people and gives space to an oppositional critique of power, based on "class solidarity" (Lindvall); fifthly, and finally, the film offers a platform to also renegotiate the persisting conflict between former Bantu and San soldiers, and the marginalization of the latter, as a consequence of the colonial "divide-and-rule" strategy. Thus, *My Heart of Darkness* implicitly addresses European colonialism's complicity and accountability of former colonial powers with regards to gross social in-equality in present-day Angola and beyond.

So far, this explorative case study has demonstrated how an analysis of the documentary film *My Heart of Darkness* in relation to its screenings, reception, and adaptation in different contexts, might indeed serve as a broader interrogation of present memory discourses related to the Angolan civil war (and beyond), social relations, and power imbalance, shifting the film's analysis towards critical cultural studies. In this view, *My Heart of Darkness* proves to be a particularly enriching subject to both memory studies and postcolonial studies and might also contribute to the promising field of postcolonial cinema studies (see Ponzanesi and Waller 2012). Due to the limits of accessible sources and further (language) barriers this study could only provide an extract of the discourse(s) emerging around the film, based on a rather limited sample. It would be especially promising to find more 'Angolan voices' and learn about their readings of the documentary and the memories it triggered. This approach thus invites a larger comparative research project dedicated to the documentary film.

Furthermore, this study also adds to Werbner's diagnosis of a "postcolonial memory crisis" and the emergence of a "critique of power" that can be observed in various post-conflict societies around the globe where marginalized articulate a "popular counter-memory" and demand recognition and social justice. As Werbner (1998, 1-2) outlines with regards to the articulation of various groups across postcolonial African states:

The demanded recognition is for a right of recountability; it is the right, especially in the face of state violence and oppression, to make a citizen's memory known, and acknowledged in the public sphere – no longer need it remain merely a private matter. In many places, people bring powerful, sometimes intimately painful traces of the colonial as well as the postcolonial past to bear on their present politics.

Here, I want to emphasize that this crisis is by far not limited to postcolonial states in Africa, such as post-conflict Angola, but strongly involves the question of accountability of 'the West' as it speaks to the legacies of European colonialism, global Cold War politics, imperial complicity in international law and Transitional Justice. Whereas, in the documentary film, Patrick and Samuel end up accusing the Angolan government, and Patrick acknowledges only the role of South Africa and Israel in the war, the involvement of other global forces like China and the US, who amongst others supported UNITA and the MPLA, remains rather absent from the discussion. As postcolonial legal scholars, who focus on the colonial legacies of present asymmetries in the distribution of power, have pointed out, 'The West' has in fact been reluctant to acknowledge its role and the imperial complicity of international law which also leads to a general suspicion towards the current TJ discourse. These legacies, however, are crucial in the realm of memory discourses and politics, such as those implied in the course of transitional justice processes and staged in the reconciliation scenario which currently seems to be gaining some dominance in the context of post-conflict Angola, as the discussion in this chapter has emphasized.

Through a 'postcolonial lens', I submit that the challenge is not only to include such critique but to constantly acknowledge and carefully scrutinize relations of power and domination within the given country as well as in international constellations which have need to be constantly put in question as TJ processes tend to exclude particular groups and disarticulate their demands. In this regard, *My Heart of Darkness* might eventually have had the potential to launch an emancipative discourse and provide a platform for dialogue about the war and its traumata, in context of a national environment dominated by amnesty laws and elite-driven memory politics; such critique of power might also address the international dimension of the conflict, which remains rather silenced in the texts analyzed in this study. A sincere discussion about accountability or "recountability" (Werbner) not only of the direct former war parties and the Angolan political elites but also of the foreign global powers would eventually enable (international) public recognition of the victims of the Angolan civil war, the Angolan people, and thus mark a starting point for larger solidarity movements and quests for social justice, both symbolically public recognition and redistribution with regards to gross social (in)equality. Such a public debate about the Angolan conflict (as in many other cases of postcolonial states), however, is still pending and arguably carefully avoided by local, national and global elites. It is in this urgent sense that I hope this study might inspire further research, forstering the study of popular culture (film) and its emancipative power as a critical intervention into political culture and memory politics.

REFERENCES

Film Credits

My Heart of Darkness – Sweden/Germany, 2010, 93 mins, directed by Staffan Julén and Marius van Niekerk. Coproduction of Gebrüder Beetz Filmproduktion, Eden Films, SVT and ZDF/arte. Supported by the Swedish Film Institute and Angolan ministry for Culture. (purchased via Journey man films)

Quoted Articles, Interviews etc.

Daniel Lindvall, "My Heart of Darkness (Sweden, 2011)", *Film International*, April 7, 2011, accessed December 8, 2015, http://filmint.nu/?p=1561

Daniel James Scott, "Director Marius Van Neikerk on My Heart of Darkness", *Filmmaker Magazine*, February 11, 2012, accessed December 8, 2015, http://filmmakermagazine.com/40348-my-heart-of-darkness-an-interview-with-marius-van-niekerk/

"Mein Herz der Finsternis", Gebrüder Beetz Filmproduktion, accessed December 8, 2015 http://www.gebrueder-beetz.de/produktionen/mein-herz-der-finsternis

"My Heart of Darkness as Tri Continental Fest", Screen Africa, August 17, 2011, accessed December 8, 2015 http://www.screenafrica.com/page/news/festivals/900964-My-Heart-of-Darkness-at-Tri-Continental-Fest#.VmlYGV6bF_s

"My Heart of Darkness", *German Documentaries*, accessed December 8, 2015 http://german-documentaries.de/films/41596

"Filme de Tony Nguxi exibido em Novembro", *Jornal de Angola*, April 19, 2011, accessed December 8, 2015 http://jornaldeangola.sapo.ao/cultura/filmes/filme_de_tony_nguxi_exibido_em_novembro

"filme-documentário Angolano: My Heart of Darkness será exibido no festival de cinema", *Platina Line – A maior revista de entretenimento de Angola*, April 16, 2011, accessed December 8, 2015, http://platinaline.com/index.php/agenda/item/1393-filme-document%C3%A1rio-angolano-my-heart-of-darkness-ser%C3%A1-exibido-no-festival-de-cinema/1393-filme-document%C3%A1rio-angolano-my-heart-of-darkness-ser%C3%A1-exibido-no-festival-de-cinema

"My Heart of Darkness será exibido no festival de cinema", *Angop – Agência Angola Press*, April 15, 2011, accessed December 8, 2015, http://www.portalangop.co.ao/angola/pt_pt/noticias/lazer-e-cultura/2011/3/15/Heart-Darkness-sera-exibido-festival-cinema,5c2eebda-dd1b-4182-b61a-ffe18820ef1f.html

"Suécia/Angola: Exibição cinematográfica marca celebrações do Dia da paz", *Boletim de Informação* n°47, Angola's Embassy, March/April 2011, accessed December 8, 2015 http://www.angolaemb.com/informacao/boletins/boletins2011/boletim-47/n22.htm

Literature

Ameida, Cláudia, Edalina Sanches and Filipa A. Raimundo. 2010. "'Bringing fighters together' A comparative study of peacebuilding and transitional justice in Angola and Mozambique." Conference paper, presented at *Working seminar Legacies of Conflict, Decolonisation and the Cold War*, LSE IDEAS. 28-29 May.

Becker, Heike. 2011. "Beyond Trauma: New Perspectives on the Politics of Memory in East and Southern Africa." (Review Article), *African Studies* 70, 2: 321-335.

Buikema, Rosemarie. 2012. "Performing dialogical truth and transitional justice: The role of art in the becoming post-apartheid of South Africa" In *Memory Studies* 5, 3: 282-292.

Casertano, Stefano. 2013. *Our Land, Our Oil! Natural Resources, Local Nationalism, and Violent Secession*. Wiesbaden: Springer VS. (Especially chapter 6 "Angola and Nigeria: tribal fragmentation in Sub-Sahara africa", 217-252).

Dubow, Saul. 2014. *Apartheid, 1948-1994*. Oxford: Oxford Histories.

Erll, Astrid and Stephanie Wodianka, eds. 2008. Introduction to *Film und kulturelle Erinnerung. Plurimediale Konstellationen*, 1-20. Berlin: De Gruyter.

Erll, Astrid. 2012. "War, Film and Collective Memory: Plurimedial Constellations." *Journal of Scandinavian Cinema* 2, 3: 231-235.

Fiske, John. 1987. *Television Culture*. London: Routledge.

Fiske, John. 1989. *Understanding Popular Culture*. London: Routledge.

Fraser, Nancy. 1997. *Justice Interruptus: Critical Reflections on the 'Postsocialist Condition'*. New York and London: Routledge.

Fraser, Nancy and Axel Honneth. 2004. *Redistribution or Recognition? A Political-Philosophical Exchange*. Translated by Joel Golb, James Ingram, and Christiane Wilke. New York and London: Verso Books.

Hall, Stuart. 1980. "Encoding/Decoding." In *Culture, Media, Language* edited by Stuart Hall, Dorothy Hobson, and Andrew Lowe and Paul Willis, 128-138, London: Hutchinson.

Hohmann, Thekla, ed, 2003. *The San and the State: Contesting Land, Development, Identity and Representation*, Cologne, Rudiger Koppe Verlag.

Koloma Beck, Teresa. 2012. *The normality of civil war: armed groups and everyday life in Angola.* Frankfurt and New York: Campus Verlag.

Malca, Camila G. and Elin Skaar. 2015. "Angola. Negative peace and autocracy in the shadow of impunity" in *After violence: transitional justice, peace, and democracy* edited by Elin Skaar, Elin, Camila G. Malca and Trin Eide, 149-173. London: Routledge.

Mathews, Sally. 2012. "White Anti-Racism in Post-Apartheid South Africa". In *Politikon: South African Journal of Political Studies* 39, 2: 171-188.

Moyo, Khanyisela. 2012. "Feminism, Postcolonial Legal Theory and Transitional Justice: A Critique of Current Trends". In *International Human Rights Review* 1, 2: 237-275.

Ponzanesi, Sandra and Marguerite Waller, eds. 2012. *Postcolonial Cinema Studies.* New York: Routledge.

Rigney, Ann. 2012. "Reconciliation and remembering – (how) does it work?" In *Memory Studies* 5, 3: 251-258.

Skaar, Elin and and José O. Serra Van-Dúnem. 2006. *Courts Under Construction. What can they do for the Poor?* CMI Working Paper 20.

Swartz, Sharlene. 2013. *The Restititution of Personhood: Exposing Possibilities for transformation through human rights films. Human Rights Research Counsil.* URL: www.hsrc.ac.za/en/research-outputs/ktree-doc/13762

Swartz, Sharlene and Duncan Scott. 2012. "Restitution: A revised paradigm for the transformation of poverty and inequality in South Africa". Paper presetend at the *Strategies to Overcome Poverty and Inequality: Towards Carnegie 3 Conference*, University of Cape Town, South Africa.

Teitel, Ruti. 2008. "Transitional Justice Globalized". In *The International Journal for Transitional Justice* 0: 1-4.

Werbner, Richard. 1998. "Beyond oblivion: confronting memory crisis" in *Memory and the Postcolony: African Anthropology and the Critique of Power*, edited by Richard Werbner, 1-17. London and New York: Zed Books.

Wilson, Richard A. 2001. *The Politics of of Truth and Reconciliation in South Africa. Legitimizing the Post-Apartheid State.* Cambridge and New York: Cambridge University Press.

Winter, Reiner. 2003. "Filmanalyse in der Perspektive der Cultural Studies". In *Film- und Fotoanalyse in der Erziehungswissenschaft. Ein Handbuch*, edited by Yvonne Ehrenspeck und Burkhard Schäffer, 151-164. Opladen: Leske + Budrich.

The Ethics of Memory in
My Heart of Darkness

ALEXANDRE DESSINGUÉ

From the very beginning, the documentary film *My Heart of Darkness* seems to have an ambitious goal and to follow a clear path. In the English trailer for *My Heart of Darkness*,[1] the ambition of the documentary is clearly and rapidly introduced by a short quotation with capital white letters on an empty black screen: "In search of reconciliation and forgiveness".[2] In this sense and aesthetically, there should be no doubt about the intention behind this cultural production. The film, which is also defined by the main character as a "frightening journey into the darkest of our hearts",[3] follows, then, a logical pattern. The story starts when the main character, Marius Van Niekerk, a South-African paratrooper, gathers three former enemies and veterans from the Angolan Civil War (1975-2002) in the middle of the Angolan bush. He invites them to recount and discuss past events with the intention of getting rid of bad memories and forgiving each other.[4] In many ways, this cultural production has an original experimental character and gives many interesting analytical research opportunities within the field of memory studies.

The main question I would like to ask in this paper is to what extent *My Heart of Darkness* actually works as a process of reconciliation and

---

1   https://www.youtube.com/watch?v=8kW9X-xTDqw, Accessed November 3, 2015
2   My Heart of Darkness, 0'38"
3   Ibid., 06'03"
4   Ibid., 03'25".

forgiveness. And if not, how can we understand and read this experimental documentary; i.e., if this film is not about reconciliation and forgiveness, what is it about?

In the following analysis, I will show that the intention behind the film is not as simple as it seems and raises several doubts about the goal(s) behind this "experimental journey". In my opinion, the ambiguity of the intention is related to the entanglement of several discourses: one discourse concerns reconciliation and forgiveness and another concerns the "banality of evil". Furthermore, another assumption correlating to the question of the ambiguous intentions behind the film is that, to some extent, it seems more motivated by Marius' individual and psychological need for self-forgiveness than by a need for a collective process of forgiveness and reconciliation.

In the second part of this paper, I will show how *My Heart of Darkness* participates in the establishment of a "new community of remembrance" and of a "new cultural memory". My second assumption is that, along with the film, we participate in a process of calibration of individual acts and voices of remembrance into a dominant collective way of telling and remembering the past, or what Avishai Margalit would call a "shared memory". [5] More precisely, I will show how this evolution towards a more authoritative way of telling the past corresponds to different discursive phases in the documentary film.

In the last part of this analysis, and as a conclusion, I will argue that because of its ambiguous goals and because of its discursive pattern, it is difficult to interpret this film as an "innocent" documentary about forgiveness and reconciliation. My third assumption is that this documentary film raises, through the re-mediation of past events and the construction of a new community of remembrance, several serious ethical dilemmas.

---

5   Avishai Margalit, The Ethics of Memory (Cambridge: Harvard University Press, 2004), 51.

THE MANY INTENTIONS OF *MY HEART OF DARKNESS*

As I already mentioned above, the first five minutes of the documentary present a much more complex picture of the intention(s) behind the film than the trailer does.

The documentary starts with a reference to Philip Zimbardo,[6] Professor of social psychology at Stanford University and known as the leader of the Stanford prison experiment: "The line between good and evil is permeable and almost anyone can be induced to cross it when pressured by situational forces. Philip Zimbardo". Without going into details, this experiment from 1971 showed how specific contextual and social circumstances could condition people to do things they would not have done in a "normal" context. One of the conclusions of the experiment was that "normal" people may become irrational and even violent under special circumstances, and that even their personality could change when the surroundings encouraged and pushed them to act in a specific way. They had to stop the Stanford experiment after only 6 days because of the psychological and physical risks it represented for the participants including for the leader of the experiment, professor Zimbardo. Similar psychological studies, such as the Milgram experiment (see Skarpeid and Wagner in this volume) from 1961,[7] had led to the same kind of conclusions raising the question of the limits between good and evil in human beings, between attitudes and acts, and the role played by authority and law in "crossing the line".

Taking into consideration that the title of the film is *My Heart of Darkness*, a clear reference to Joseph Conrad's novel, the intention of the documentary seems to open up other perspectives, at least at this stage, which are not only about reconciliation and forgiveness. The film also implicitly problematizes the question of the "banality of evil" at an ontological plan, which is a central notion in Hannah Arendt's report on the

---

[6] For more information about this experiment, see among others C. Haney, W. C. Banks, & P. G. Zimbardo, "Study of prisoners and guards in a simulated prison", Naval Research Reviews, 9, (1973): 1–17; and. C. Haney, W. C. Banks, & P. G. Zimbardo, "Interpersonal dynamics in a simulated prison", International Journal of Criminology and Penology 1, (1973): 69–97.

[7] Stanley Milgram, "Behavioral Study of Obedience", Journal of Abnormal and Social Psychology 67 (4), (1963): 371–378.

trial of Eichmann in Jerusalem (2006). In this report, Arendt argued that Eichmann wasn't a monster but an obedient administrator of the Reich and of the final solution: "He did his *duty*, as he told the police and the court over and over again; he not only obeyed *orders*, he also obeyed the law".[8] Eichmann himself was sure that he was not "a dirty bastard in the depths of his heart",[9] and that he had nothing to do with the final solution or with the killing of Jews: "I never killed a Jew, or a non-Jew, for that matter – I never killed any human being. I never gave an order to kill either a Jew or a non-Jew; I just did not do it".[10] Even the psychiatrists responsible of the psychological evaluation of Eichmann at the trial certified him as "normal".[11] Arendt seems to imply that despite his "normality" Eichmann is "devilish", and this is what "the banality of evil" is actually about. In this sense, the main ethical issue in Arendt's report raises an ontological question for all human kind, not only for a "minority of monsters". The Milgram and the Stanford experiments were based on the same assumptions and the documentary film *My Heart of Darkness* raises the same kind of ontological questions.

But the ambiguity of the intention behind the film is not only related to the question of "what is this film about?" but also to the question of "*who* is this film about?". This uncertainty is clearly expressed through Marius' own formulations in the film. According to the main character, the film asks the following: "what does a man have to do to regain his self-worth?".[12] However, it also asks how "to come to turns with my [Marius'] past, with my [Marius'] memories from the war"[13] and how "to get rid of it once of all"[14]. Later in the documentary he also talks about this journey as a "frightening journey" and adds: "We are all here for the same reason driven by the need to understand, to set off on a frightening journey, a journey into

---

8 Hannah Arendt, Eichmann in Jerusalem. A report on the banality of Evil (London: Penguin Books, 2006), 135
9 Ibid., 25.
10 Ibid., 22.
11 Ibid., 25.
12 Ibid., 01'30.
13 Ibid., 02'00.
14 Ibid., 03'25".

the darkest of our hearts".[15] The unstable use of the language and in particular of the pronouns in Marius narratives introduces a certain uncertainty about the question of who this journey is about and who it is for. The inconsistent use of the pronouns actually indicates different foci, from a very impersonal and general perspective ("a man", "his self-worth") to a much more personal and intimate perspective ("my past", "my memories") to a collective one ("we", "our hearts"). The fact is that Marius is the one who took the initiative to gather the veterans for this experimental documentary and we could legitimately ask if the intention of the film presented by Marius as a "journey into the darkest of our hearts" appears to be foremost a journey into the darkness of Marius' heart. At this stage of the film, nothing implies that it is about something other than Marius' individual journey, his personal need to rid himself of his bad memories, and the fact that to overcome them, he chooses to travel back to Angola to meet other veterans from both sides.

In my opinion this hypothesis is comforted by the way the question of responsibility and guilt is represented in the film. In the narrative of past events made by the four veterans, i.e. the South-African Marius, and the Angolans Mario, Samuel, and Patrick, there is no doubt that all four have taken part in or even been the victims of war atrocities. Once again, however, Marius seems to occupy a special position in the film, and not only because he is the one who took the initiative to organize this journey. The fact that he kept gruesome pictures of killed enemies in a shoebox (explicit traces of violence and cruelty) also reinforces his special status in comparison with the other veterans. This doesn't mean that the Angolan participants in the documentary did not commit the same kind of atrocities; the film doesn't actually thematise this topic in the case of the Angolan soldiers. But the fact that the documentary doesn't explicitly show similar evidence of the Angolan soldiers' cruelty, strengthens the spectator's impression that the responsibility mostly belonged to Marius. Even when Samuel tells the horrific story of burning innocent women as a soldier with the Savimbi army,[16] he never seems to be presented as "guilty" as is Marius. Marius seemed to act on his own, while Samuel is presented as a

---

15 My Heart of Darkness, 06'03", I underline.
16 Ibid., 42'00-45'00.

simple soldier who had to follow orders; Marius took pictures of his crimes; on the contrary, Samuel says that he cried.

The role of these photographs at the beginning of the film and the aesthetic effect they create seem to be central in the spectator's perception of the question of responsibility for these crimes: "who is the guiltiest?". Marius' photographs are full of blood, almost insupportable and particularly unjustifiable. One of the pictures represents a dead body whose face is unrecognisable, it is even difficult to see that it is human; in another picture, a South-African soldier is clearly portrayed as a hunter beside a dead body; and in a third picture, three South-African soldiers stand beside a dead body, one of the soldiers holding a foot in his hand. These pictures are clearly characterized by a morbid scenography. Even though we do not see Marius in some of these pictures, the spectator assumes he was involved as he presents them as his own. There are serious ethical issues involved in taking pictures of dead people in war scenes and then keeping these photos. The difficulty isn't only that these are pictures of dead bodies, but the way these dead persons are represented as trophies. I believe this influences the spectators' perception of the question of responsibility throughout the documentary and also their perception of the different kinds of agency each veteran actually exercises, i.e. the film asks the question "who actually had the power to act as an independent individual?". In other words, the four veterans are not portrayed equally in terms of agency in the documentary when it comes to their acts during the war. They are not depicted as sharing equal responsibility for the "banality of evil".

In this sense, it is difficult to perceive this experimental journey as the result of a clear and equal need for all four participants. One question we could ask at this stage is to what extent all four participants are similarly conscious of the experimental and ontological character of the film. Marius asserts that they are all here for the same reason, but they are presented as having different responsibilities, roles, and attitudes in relation to the events and atrocities committed during the war. This experimental documentary seems to be clearly motivated by a personal need, not a collective one. In the next part of this paper, the analysis of the different discursive phases in the film will confirm the fact that the personal ambition(s) of Marius will slowly influence the other veterans. Conducting them to some form of acceptance or resignation, the documentary will end with the construction of some kind of shared intention. It is the narrative evolution of this process that I will attempt to analyse now.

FROM A COMMON MEMORY TO A SHARED MEMORY

My main assumption in this part of the analysis is that, through viewing *My Heart of Darkness*, the spectator assists step-by-step in the construction of a new cultural memory about the Angolan Civil War. This memory is not to be considered as a primary act of remembrance but is shaped by the re-mediation of past events through the narrative evolution of the film. This means that an interesting aspect of this documentary not only resides in its character as a cultural memory (a new cultural production about the Angolan Civil War), but also in its experimental character, i.e. its narrative evolution demonstrating how the construction of a new cultural memory occurs. More precisely, based on the notions introduced by Avishai Margalit in *The Ethics of Memory*, I will show that the construction of a new cultural memory in *My Heart of Darkness* depends on the way the memories are narrated, mediated and re-mediated alongside the film, first framed within the concept of "common memory" and then evolving into a concept of "shared memory".

A common memory is defined by Margalit as an "aggregate notion": "It aggregates the memories of all those people who remember a certain episode which each of them experienced individually".[17] The common memory of the Angolan Civil War represented in the film (i.e. the episode according to Margalit), could actually be understood as the result of the aggregation of the different individual testimonies made by the four veterans (see Fig. 1).

---

17 Ibid.

Fig. 1. The first discursive phase in the documentary film, which presents the Angolan War as a common memory.

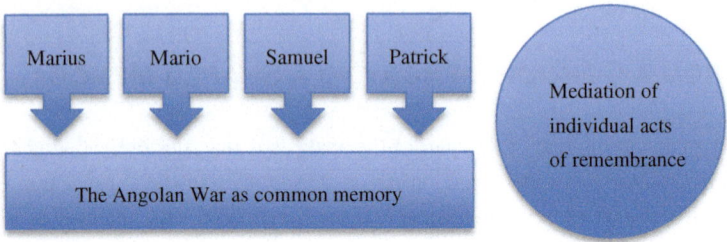

The term "common memory" is a bit confusing (as is the Halbwachian term of collective memory)[18] since it can also be interpreted as "unified memory", which is not in Margalit's definition. The meaning of the term "common memory" has to be considered in relation to individual acts of remembering. A common memory isn't the result of an addition but of an *aggregation* of individual acts of remembering. In that sense, a common memory can be interpreted as preserving a certain polyphonic nature, because the voices of the individuals don't merge in a syntagmatic narrative. This aspect is clearly identifiable at the beginning of the film.

My Heart of Darkness begins by mediating a common memory of the Angolan Civil War from the perspective of different individual narratives. To start with, all four of the veterans share their experiences of the Angolan Civil War from an individual point of view and through more or less personal lived memories.

Patrick, an Angolan soldier recruited at the age of 14 by the MLPA,[19] starts by stating "I don't have anything bad to say about the war [...] No one was bad to me. Everyone was good to me. [...] the war was good",[20] and continues arguing that his participation in the war was a necessity:

---

18 For further discussion see Alexandre Dessingué, "From Collectivity to Collectiveness: Reflections (with Halbwachs and Bakhtin) on the Concept of Collective Memory" in Siobhan Kattago, The Ashgate Research Companion to Memory Studies. (Dorchester: Ashgate, 2015).

19 The People's Movement for the Liberation of Angola (MPLA) was supported by the Soviet Union and by Cuba.

20 My Heart of Darkness, 11'00-11'14.

"But you South Africans invaded Angola with the Israelis, with superior power [...] No I couldn't sit still. I was forced to fight against you.".[21] In this narrative, the voice of Patrick is clearly identifiable; he expresses both personal feelings and ideologically influenced conceptions of the Angolan Civil War.

The same happens to Samuel who was recruited by UNITA[22] at the age of 15. Samuel remembers the war through the first battle in which he participated: "The first bullet from the enemy hit my friend straight in the head";[23] then he recounts the life conditions during the war:

In war, you don't sleep. You don't take off your pants or your shirt, you sleep like a goat in the bush. You know that at any moment, your enemy can attack and if you don't run you are dead.[24]

In this case, Samuel's witness of the Angolan Civil War is personal and rooted in an individual memory, even if later his witnessing acquires a more general character.

Mario, recruited by the Portuguese (colonial power in Angola) at the age of 15, chooses mostly to share his experience by referring to contextual factors. He doesn't really tell us individual memories of the conflict or of the war; the use of the "we/us", and not the "I/me", is characteristic of Mario's narrative:

The Portuguese, Namibian and the South African army all exploited us [...] they used the bad relation between the Bushman and the Bantu [...] The Portuguese come to us with false pretentions and misused us for their own objectives.[25]

While Samuel's story is more personal and intimate, Patrick creates a mixed narrative of both personal memories and contextual, ideological explanations. Mario chooses to never comment on any personal memories

---

21　Ibid., 17'26-17'36"
22　The National Union for the Total Independence of Angola (UNITA) get support by the USA, China and South Africa.
23　My Heart of Darkness, 11'35"
24　Ibid., 11'38"
25　Ibid., 14'58"-16'00

of the war. He doesn't actually share any individual, concrete experiences, instead making more general comments about collective historical considerations and collective acts of remembrance.

In his own telling, Marius explains first that it was natural for him to become a soldier in South Africa, and that it was a question of education, religious education and of the "defence of values".[26] Then he starts describing his first war operation as a young soldier.[27] At this stage of the documentary, the veterans clearly have different approaches and have chosen different ways of witnessing the war. These acts of telling and remembering are different both in their content (what they tell) and in their form (how personally/contextually they are formulated). In this first part of the documentary, all four veterans present a memory of a common episode (the Angolan Civil War) from different individual and personal points of view. At this stage, there is no trace of any integration of the individual perspectives towards a more "holistic" narrated memory.

Furthermore, the three Angolan veterans, and more specifically Patrick, are very distant from Marius' story during the first thirty seven minutes of the documentary. On several occasions, they even contest the fact that they are sharing a common experience of the past with him. This is the case when Marius starts telling the story of a never ending nightmare: "The guy that I killed, his face appears in different ways in dream to me".[28] Through this story, he tries to explain his individual traumatic experience, but the incomprehension from Mario is manifest and the reaction of Patrick is straight: "I don't want you to tell us about your dreams, I want real things... I don't believe in dreams".[29] Later on, when Marius says that he wants to show them pictures from the war, the reaction from Patrick is again immediate: "No, we don't need it"[30] and a bit later he comments ironically and critically: "He had a good life, being able to take pictures".[31] When Marius explains that he wants to burn the pictures and that he wants

---

26 Ibid., 19'15"-20'00"
27 Ibid., 20'18"
28 Ibid., 29'57"
29 Ibid., 32'28"
30 Ibid., 34'00
31 Ibid., 34'26"

to get rid of that memory,³² Patrick's reaction is again a clear protest: "For me, I don't trust pictures.".³³ At that particular moment, it really seems like the Angolan veterans don't really understand what the journey is about. They seem to neither share nor want to share their experience of the past in an intimate way.

Nevertheless, the spectator will assist throughout the documentary in a step-by-step evolution towards a re-mediated and culturally constructed memory defined by Margalit as a "shared memory":

A shared memory [...] is not a simple aggregate of individual memories. It requires communication. A shared memory integrates and calibrates the different perspectives of those who remember the episode [...]³⁴

A shared memory introduces a holistic and syntagmatic perspective on individual acts of remembering through what Margalit calls integration and calibration (according to his definition introduced above). A shared memory does not result from the addition of individual memories but from their calibration into particular "mnemonic devices",³⁵ in this particular case, the documentary film. In this sense, and as I understand it, a shared memory is closer to the notion of cultural memory,³⁶ because it works as the result of a cultural representation and selection, as "a second order memory – or a memory of a memory (or memories)",³⁷ or as Margalit puts it: "A shared memory of a historical event that goes beyond the experience of anyone alive is a memory of a memory [...] This kind of memory reaches alleged memories of the past but not necessarily past events.".³⁸

In my opinion, the evolution through the documentary from a common memory to a shared memory starts precisely after Marius has finished

---

32 Ibid., 35'04
33 Ibid., 37'28"
34 Margalit, The Ethics of Memory, 51.
35 Ibid., 54
36 Jan Assmann, "Communicative and Cultural Memory" in A. Erll & A. Nünning, Cultural Memory Studies, (Berlin: Walter de Gruyter, 2008), 109-118.
37 Alexandre Dessingué, A. & Jay M. Winter, Beyond Memory: Silence and the aesthetics of remembrance. (London: Routledge, 2016), 4.
38 Margalit, The Ethics of Memory, 59.

talking about his own experiences of the war and concludes by asking in a very impersonal way: "How is it possible to drag such baggage around?".[39] In this way, Marius starts the process of memory calibration and the integration of different individual memories into a shared experience and narrative. It is, I believe, at that particular moment, that the process starts evolving from a common memory, i.e. an aggregation of individual memories, to a shared memory, i.e. an integration and re-mediation of individual memories in a collective narrative about the past.

This evolution in the narrative is conspicuous when Marius expresses more personal feelings directly to the other veterans: "I'm feeling very hurt about you guys not trusting me."[40] It is also at this very peculiar moment that Mario stresses the topic of forgiveness for the very first time: "We just said that the way you (Marius) want forgiveness is difficult for us."[41] The question of this topic is central in the documentary because it implicitly introduces the possibility of creating a narrative within a shared frame. At the same time, the dialectical relations between the You (Marius) and We (the three Angolan veterans) will weaken and will give way to the establishment of a stronger collective "We" constituted by the all four veterans together:

Marius: I feel the need to apologize, to get forgiveness from Angola.
Patrick: You need to apologize. It's your right. Because you came to Angola to fight. And me, I was defending myself. [...]
Marius: I feel that I would like to ask for forgiveness for what we've done. Don't Samuel and Mario feel remorse or feel that they need forgiveness from the Angolans [...]?
Mario: What's he asking?
Patrick: If you personally want to apologize to your people.
Samuel: But where shall I find them, and where shall I apologize? We can ask forgiveness here. We left all of that behind us.
Patrick: What should be done isn't an apology from Samuel but an apology to the people. What I did was wrong.

---

39 My Heart of Darkness, 27'05"
40 Ibid., 37'30"
41 Ibid., 39'15"

Mario: What makes you want to ask for forgiveness? It's probably that you feel bad, that you feel deceived. [...] but I am here to ask forgiveness from the Angolans, our brothers.
Patrick: What Marius is trying to do... For me, it's very important. For me it's very important. I know we need that, too. Marius also needs... We're going to help you to do that.
Marius: Together. I wouldn't be able to do this without your support and recognition. And that is... I have it, and it makes my heart feel very good. Thanks a lot.
Patrick: I'll give my whole support to you.
Marius: And I'll support you too, Patrick. We'll support each other.
[Samuel and Mario agree, they nod and confirm that they will also support each other.][42]

My opinion is that this dialog plays a major role in the documentary, which starts with a clear opposition between Marius and Patrick: "I feel" vs. "You need to". But right after that, Marius suggests a new shared experience with Samuel and Mario when he asks them: "Don't Samuel and Mario feel remorse or feel that they need forgiveness from the Angolans [...]?". As I already mentioned, the topic of forgiveness is a key notion in the documentary because it works as a kind of catalyst or a calibrating frame. At this particular moment, the documentary evolves from focusing on the needs of a white man who wants "to get rid of his bad memories" to focusing on a shared experience: "asking for forgiveness to the Angolan people". Through the particular question formulated by Marius and the affirmations of Mario and Samuel, the documentary draws a new collective memory accompanied by the construction of a new "collectivity of remembrance".[43] At the same time, this dialog establishes a new dialectical relationship among the group. On the one side, we have the ones who have to ask for forgiveness from the Angolan people; on the other side, we have Patrick who still considers that he fought on "the right side," i. e., for the Angolan people.

This dialectical relationship will also change rapidly when Patrick himself unexpectedly recognizes the importance of the process Marius is

---

42 Ibid., 39'22" – 41'53"
43 Dessingué, "From Collectivity to Collectiveness", 100.

trying to carry out: "What Marius is trying to do... For me, it's very important. For me it's very important. I know we need that, too.". Here Patrick becomes a part of the process of forgiveness itself and integrates into the new established collective "we". Actually, the rapid evolution of the film at this stage is surprising, and we may not understand why and how the narrative has developed as it has, especially when Marius says that he will support Patrick. The reason why we may be surprised is that until then Patrick has never expressed a real need for "being supported"; but the internal dynamics of the conversation between Marius and Patrick clearly lead to the establishment of a stronger "we" through the establishment of a shared intention and narrative. The personal motivations and individual voices have, in this sense, merged into a shared narrative.

At this stage, we can assume that individual acts of remembrance are turned into a collective shared narrative about the past motivated by a shared intention and calibrated by the topic of forgiveness. Through the re-mediation of the individual memories into a shared memory, the spectator actually witnesses the creation of a cultural memory. This process is even concluded by the ritual at the end of the documentary, where the photos from Marius are burned, and where all four veterans become "official" members of a new collectivity of remembrance. During this evolutive process from a common to a shared memory, we experience in *My Heart of Darkness* what Astrid Erll calls the basic functions of cultural memory: "they make the past intelligible; and [...] they play a decisive role in stabilizing certain mnemonic contents into powerful sites of memory".[44] In this sense, the topic of forgiveness functions as a catalyst for re-mediation, making the past events intelligible for the four veterans and the ritual at the end functions as a stabilizer of memory or as a new site of memory. In many ways, this documentary film contributes to the creation of a new cultural memory not only for the spectator but also for the veterans, i.e. a new stabilized mnemonic devise which has calibrated, step-by-step, the veterans' multiple voices and experiences into a shared cultural memory (see Fig. 2).

---

44 Astrid Erll, Memory in Culture, (Basingstoke: Palgrave Macmillan, 2011), 143.

Fig. 2. Second discursive phase in the documentary film, which leads to a new collective remembrance of the Angolan War as a shared cultural memory.

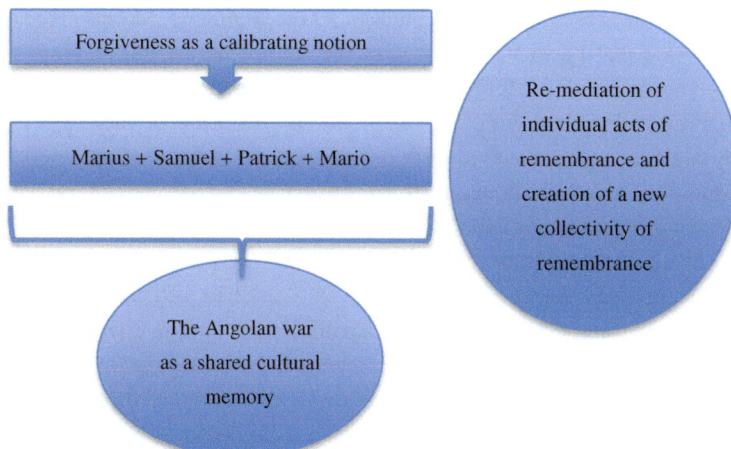

CONCLUDING REMARKS: FORGIVENESS AND/OR FORGETTING?

My third assumption is that the re-mediation of these individual acts of remembrance into the creation of a shared memory contributes also to the shaping of several ethical dilemmas. I would like to problematize these in the final section of the paper.

In *The Ethics of Memory*, Avishai Margalit makes a clear difference between what he calls forgiveness as a "covering-up model" and forgiveness as a "blotting-out model".[45] Forgiveness, according to Margalit, includes a necessary process of forgetting, but the "blotting-out model" includes an obligation of forgetting (for example, an amnesty law), while the "covering-up model" integrates an acceptance of forgetting *by the victims*. Since *My Heart of Darkness* is introduced as a documentary film about forgiveness and reconciliation and since the main character stresses the fact that they have to forgive each other, an interesting question raised

---

45 Margalit, The Ethics of Memory, 203.

in this last section would be "what kind of forgiveness is this documentary film about?".

As the previous analysis has shown, many elements of the film lead us to think about this process of forgiveness as an initiative forced by the main character, i.e. as a blotting-out process. At the beginning, Marius is the one who says that he wants to get rid of the bad memories and he is the one who needs to ask for forgiveness. The other veterans don't share the same ambitions, but they "will get there" through the film or more precisely through the experimental journey they share with Marius. In this sense, the film raises an ethical dilemma: to what extent is this process of forgiveness driven by the veterans and to what extent is it actually imposed upon them?

Paul Ricoeur argues that any acts of forgiveness presuppose the individual and direct recognition of an identified offender and of an identified victim.[46] And this poses another clear ethical dilemma in this documentary. In the case of *My Heart of Darkness*, the offenders are identified, but the victims are not explicitly represented. One of my assumptions is that the protagonists of the film cannot forgive each other because there is no clear representation of a victim-offender relationship between them. Furthermore, this documentary cannot be about what Ricoeur calls forgiveness as "an ethical use of forgetting",[47] because it depends on an individual, active, and voluntary *oversight* of a past event. The fact that several times the veterans talk about the necessity of asking for forgiveness from the "all Angolan people" traduces the absence of the representation of the victims and, in this respect following the reasoning of Paul Ricoeur, it also traduces the unrealistic character of the process of forgiveness and reconciliation promoted by *My Heart of Darkness*.

Forgiveness and forgetting are mainly used in the documentary as an "existential notion", a kind of "self-forgiveness" or "self-worth", as both Marius and Mario discuss. As Margalit put it, forgiveness becomes a necessity "for the sake of carrying on".[48]

The main ethical dilemma of this documentary is actually linked to the attempt to make an individual journey a collective one. In this process, as

---

46 Paul Ricoeur, La Mémoire, l'histoire et l'oubli. (Paris: Editions du Seuil, 2000), 585-589, my translation
47 Ibid.
48 Margalit, The Ethics of Memory, 208.

we have seen, the role of Marius is central. But it is also important to note the fact that all four veterans agree to participate *in a common and then a shared* experimental journey. The most explicit and interesting result of this experimental journey is the evolutive and creative process of turning a common memory into a shared memory. This happens through the interaction of the participants, and also because the film is objectified as cultural memory. In this case, we may conclude that the process of creation (the documentary) doesn't lead primarily to an act of forgiveness or reconciliation but, somehow, to an act of collective re-mediation and collective oversight. More precisely the documentary film participates in the partial replacement (oversight) of a common memory at an individual level by a new narrative based on a shared narrative about the past and the establishment of a new collectivity, understood as a community of remembrance. In this sense, it is interesting to consider the notion of the community of remembrance in the light of Lave and Wenger's "community of practice", defined as a community where members are involved in a set of relationships over time.[49] Through this experimental journey, all four veterans progressively acquire a certain sense of joint identity that also involves multiple ways of sharing things, ideas and concepts. The construction of the shared memory that we assist through this experimental journey shows to what extent this process is a dynamic one, consisting of complex discursive interactions and processes, but it also clearly problematizes the question of the power and dominance in such relations, revealing clear ethical dilemmas.

---

49 Jean Lave and Etienne Wenger, Situated learning. Legitimate peripheral participation, (Cambridge: University of Cambridge Press, 1991), 98

REFERENCES

Arendt, Hannah. *Eichmann in Jerusalem. A report on the banality of Evil.* (London: Penguin Books, 2006).

Assmann, Jan. "Communicative and Cultural Memory" in A. Erll & A. Nünning, *Cultural Memory Studies*, (Berlin: Walter de Gruyter, 2008): 109-118.

Dessingué, Alexandre & Winter, Jay M. *Beyond Memory: Silence and the aesthetics of remembrance.* (London: Routledge, 2016).

Dessingué, Alexandre. "From Collectivity to Collectiveness: Reflections (with Halbwachs and Bakhtin) on the Concept of Collective Memory" in Siobhan Kattago, *The Ashgate Research Companion to Memory Studies.* (Dorchester: Ashgate, 2015), 89-102.

Erll, Astrid, *Memory in Culture*, (Basingstoke: Palgrave Macmillan, 2011).

Haney, C., Banks, W. C., & Zimbardo, P. G., "Interpersonal dynamics in a simulated prison", *International Journal of Criminology and Penology* 1, (1973): 69-97.

Haney, C., Banks, W. C., & Zimbardo, P. G., "Study of prisoners and guards in a simulated prison", *Naval Research Reviews*, 9, (1973): 1-17

Lave, Jean and Wenger, Etienne, *Situated learning. Legitimate peripheral participation*, (Cambridge: University of Cambridge Press, 1991)

Margalit, Avishai. *The Ethics of Memory*, (Cambridge: Harvard University Press, 2002).

Milgram, Stanley. "Behavioral Study of Obedience", *Journal of Abnormal and Social Psychology* 67 (4), (1963): 371-378.

Ricœur, Paul. *La Mémoire, l'histoire, l'oubli.* (Paris: Editions du Seuil, 2000).

Memory, Contradictions and Resignification of Colonial Imagery in *My Heart of Darkness*

STEFFI HOBUß

Different scholars in memory studies hold that memory does not simply refer to the past but always deals with the present. Time and again, media and cultural archives allow and require actualizations of the remembered past and – perhaps – even forgotten or taboo things. Social memory produces present constructions of the past. How do the actualizations work in the case of the documentary *My Heart of Darkness*?

In the first section, I will present an introduction to the idea of collective memory understood in terms of present memory acts. Against this backdrop, two further sections will examine contradictions as a central part of the film's aesthetic and give some examples of resignification and remediation of colonialist and racist imagery. *My Heart of Darkness* is a documentary about and entrapped by contradiction, e.g. on the level of dialogue, and in the role of the photographs and other material things presented. The main character, Marius van Niekerk, wants to get rid of his photos of the Angolan civil war by burning them. This is his individual intention as shown in the film. But in the making of the film, the photos are preserved; the fact of making the film itself serves to safeguard them. Thus, more than the character's expressed individual intention is shown; Marius' wish for private ownership of his memories is a desire impossible to fulfill. Remediations and resignifications are another important issue. Considering the series of connections with Joseph Conrad's *Heart of Darkness* via Coppola's *Apocalypse Now* to Julén's/Van Niekerk's *My Heart of*

Darkness, what are the results of the new revisionings? The issues of colonialism and racism seem to be swallowed up by the "universal" aim of forgiveness; thus, one could read some color blindness at work in the documentary. Therefore, the third part will give attention to the uses of daylight, firelight, and darkness, and to the pictures of animals and nature that are connected with colonial imagery.

My Heart of Darkness, with its contradictions and with its gap between the – perhaps – naïve intention and its possible interpretations, seems to be a documentary about the contradictions of memory, about the aesthetics and the present possibilities of remembering, and about how something is remembered (through a photo, a movie, a documentary), rather than a film "about the war".

AGAINST "REPRESENTATION" – COLLECTIVE MEMORY UNDERSTOOD IN TERMS OF PRESENT MEMORY ACTS

Many theories and interpretations in memory studies use the concept of "representation". Sometimes "representation" seeks to describe the structure of remembering the past while in other cases objects and artifacts are classified as representations in memory studies. I, however, consider the use of this concept as unfavourable for several reasons and suggest thinking about memory without representation. From my perspective, the uses of the concept of representation always leads to dualism and looking back.[1]

A passage from Wittgenstein's *Philosophical Investigations* (Wittgenstein, 2001: 518-522)[2] will provide arguments against reference and representation in the context of memory and explain how these concepts lead to dualism. In other places, I have been using Wittgenstein's so called "private language argument" (PI 258) to illustrate his view towards memory: There is no such thing as a private, inner foundation of

---

[1] Some of these thoughts have been developed in discussions with John Sundholm.

[2] As is customary, Wittgenstein's Philosophical Investigations will be quoted by the acronym PI followed by the number of paragraph rather than the page.

memory (Hobuß, 2010 and 2013). But in the case of our present discussion, it is worth looking at another section, PI 518. Here, Wittgenstein deals with referential and non-referential pictures. Wittgenstein's philosophy of language compares sentences, utterances and texts to pictures in manifold ways. Wittgenstein examines the consequences of understanding texts and language as images or copies, and of comparing pictures to sentences. (For good reason, he does not invent a differentiation between linguistic and pictorial modi.) PI 518 reads as follows:

Socrates to Theaetetus: 'And if someone thinks, mustn't he think something?' –Th.: 'Yes, he must.' –Soc.: 'And if he thinks something, mustn't it be something real?' –Th.: 'Apparently.'
   And mustn't someone who is painting be painting something – and someone who is painting something be painting something real! – Well, tell me what the object of a painting is: the picture of a man (e.g.), or the man that the picture portrays?

This section belongs to a context dealing with the question whether the meaning of words and sentences can be established by means of mental pictures or imaginations. Such a theory stands for a kind of a theory of reference, in this case assuming that words and sentences are meaningful if they are correlated with certain imaginations or mental ideas. Therefore we can talk about absent things because 'inner' or 'mental' imaginations or images represent the object of reference; Wittgenstein rejects all those theories. In the second paragraph, after quoting the passage from the Socratic dialogue, he puts the same question to painting – he shifts from imagining something to the painting of something. In this case, we can ask: What is to be called the object of a painting?[3] What sort of something? Wittgenstein points at the fact that this something can be understood by a twofold interpretation, because the question has at least two answers: If the act of painting creates "the picture of a man", the material painting that can be hung at a wall e.g., then it indeed produces 'something', but this need not be an image of something real. Or its object is "the man that the picture portrays"; the act of painting leads to a depiction, the object being the man

---

3   Important: The German original has "das Objekt des Malens", i.e. "the object of (the act of) painting".

outside the picture. Significantly, Wittgenstein does not seek a definite and unambiguous paradigmatic meaning of what is to be understood by "the object of painting"; rather, he shows the different possibilities of understanding the expression. He rejects the idea that imagining/picturing always has to be the same thing or act; he also rejects the idea that all cases of speaking-about something always have to follow the same structure. In the case of the pictorial imagination in PI 518, it is difficult to determine what is to be understood by "the picture". This leads to the insight that in the case of imaginations, the use of the concept of the image or of picturing raises difficulties. The fact that there is no distinct and unequivocal interpretation of the "something" shows that the concept of the picture is highly problematic when used in this context. And when we are looking at a picture only, we cannot see whether it is a picture in the sense of the portrait of a real person or a picture without a reference to an external person. But because it is a picture we are inclined to forget that the "something" need not be anything outside of the picture itself. This is a critique of our practices and interactions with pictures: Wittgenstein shows that we are seducible by pictures – seducible to think that there is "something" the picture "represents". Wittgenstein constantly rejects the idea that there are characteristics in the structure of pictures or sentences themselves that tell us whether the picture, the sentence, or the text has a meaningful use. It is not by means of the structure or the characteristics of a state of facts that we understand it as meaningful or senseless. Sentences, texts, and other meaningful objects can be compared to images but this comparison can be ambiguous and/or vague. And thence there is no such thing as the alleged homogeneous relation of representation between a state of facts and its image.

When we apply these thoughts to memory studies, the concept of representation is misleading because it leads to the idea of a homogeneous relationship between the object or artifact and "something" that is remembered. Thus it brings about a false unification of all different cases of remembering, moreover it is an idea connected with dualism. Implicitly, Wittgenstein gives us a hint by using of all philosophical texts the quotation from Plato's *Theaitetos*: Usually in Plato's dialogues, Socrates is trying to find the unified meaning of a concept. By choosing this quotation, Wittgenstein emphasizes that the ideas he deals with have been important and influential since the time of Plato. The concept of representation might be a Platonist one and as such deep-rooted in our thinking. Socratic unification is in many cases seemingly helpful because it reduces complexity; its dualism is seemingly helpful because it seems to explain meaning by reference to something else. But above all, concentrating on the "something" that is remembered directs us towards the past; thus, using the concept of representation in the context of memory studies is past-oriented, and it will be helpful to follow Wittgenstein and stop using the term "representation" because of the seductive force of that word.

There follows another result: No pictures or texts or films and their elements are on their own politically helpful, socio-critical or mere affirmations; it always depends on their readings, their interpretations, and their uses in the Wittgensteinian sense that allows them to gain affirmative or subversive force.[4] Nevertheless, in certain contexts, certain readings or uses are made possible or are offered, and thus they are – in opposition to some objections – not completely arbitrary. Among these contexts are conventional rules of reading and picturing, but also historical and societal positions of speakers and agents.[5] This holds for memory practices as well. Memory and the processes of remembering cannot be controlled by an autonomously and voluntarily acting subject, nor by a representational connection or reference to history-as-it-was; rather, they are negotiated, consensual or not, sometimes even violently or in forms of war. This does not weaken the category of responsibility; to the contrary, it stresses the important role everybody takes in these processes of negotiation.

Following Austin's theory of speech acts, I suggest the notion of memory acts (Austin, 1975; for a more detailed explanation see Hobuß 2013). As Austin developed his theory, he changed it in an important way. Initially he suggested the distinction between constative and performative utterances: constative utterances give factual descriptions of the world and

---

4   See Kaya de Wolff's paper in this volume: She reads My Heart of Darkness as an ambivalent cultural text that allows very different readings.

5   The validity of such conventions is explained by the theory of performativity according to Austin, Butler und Derrida: The standing of conventions presupposes their repeated performance again and again, and there are always failures, miscarriages and shifts of meaning. Herein lies the theoretical fundamentum of the possibility of change. The third part will return to the possibilities of resignification and to speaker's positions.

are either true or false, while performative utterances are acts that do not describe the world, but do something by speaking, e.g. make a promise or cry out for help. But then Austin came to notice that utterances cannot simply be divided by a sharp distinction between cases of constative utterances on the one hand and performative utterances on the other. He changed his theory and stopped looking for different *cases*, but for different *aspects* of utterances. The case distinction has been replaced by a distinction of aspects. A single utterance can have a descriptive force and an acting, performative force at the same time. We can understand Austin's change as a methodological paradigm for memory studies and ask: How are memory acts carried out in specific contexts? This accounts for memory practices and their very different readings in a non-dualist, present-oriented way.

What memory acts can be found in *My Heart of Darkness*, then?[6] A memory act can be carried out by

1. illustrating[7] a past event or process (by linguistic, nonverbal, pictorial, aesthetic, scientific or other means) as an individual, e.g. talking about past experiences. Plenty of examples can be found in *My Heart of Darkness*: immediately in the opening sequence, Marius talks about how difficult it is to "regain worth" after having lost his family to exile, his sanity to trauma, and his innocence to war; he also discusses his need to come to terms with his past and his memory (Julén/Niekerk 2010, 00:33-2:05), as well as his time at school in South Africa and his entry into the army (Julén/Niekerk 2010, 18:29ff.); finally, there is the scene when the others – Mario, Samuel and Patrick – talk about their recruitment at the age of forteen or fifteen (Julén/Niekerk 2010, 11:05; 12:35; 14:44).

---

6   Memory acts can be very different things; the following list is of course open and to be completed.

7   This is the place where the notion of "representation" is often used. But it is better to avoid the word and the misunderstanding as 'simply' copying something original that was there before that act of representation. Rather it is to be understood as memory work or process that in some cases illustrates something that exists only in the specific means of illustration. For further details about representation, see: Steffi Hobuß (2012) (In this article I tried as an experiment to save the concept of representation.)

2. illustrating a past event or process (by linguistic, nonverbal, pictorial, aesthetic, or other means) as a member of a specific group. There need not be a sharp boundary between these first two aspects. For example, after Mario reported his recruitment by the Portuguese at the age of 15, he changes his perspective slightly, but important, and says "The whites used us, threw us away and today nobody cares about us" (Julén/Niekerk 2010, 15:54). Here, a we/they-distinction between Black and White people is built, and Mario accuses the Whites as a member of a Black "we".

3. claiming that something should be remembered by others (another individual person or a representing group), and

4. sharing memories with others. This seems to be one of the main aspects of *My Heart of Darkness* and is connected with the issues of forgiveness and reconciliation. The whole documentary is plotted as a journey of "me and three of my former enemies", as Marius puts it (Julén/Niekerk 2010, 5:09), and "what we share lies ahead of us", namely "that battlefield that destroyed us"(Julén/Niekerk 2010, 7:39). There are memories that are to be shared, while "ahead of" can be understood twofold: In terms of space, they are travelling into the region where they fought against each other during the war, and in terms of time, they are shown in a process of sharing their memories more and more. In the beginning, especially Patrick is shown as skeptical: "I don't have anything bad to say about the war" (Julén/Niekerk 2010, 10:51); later, in a watershed scene of the film, Mario says "the way you want forgiveness is difficult for us" (Julén/Niekerk 2010, 39:14). But then, they reassure each other that they will support and help each other in sharing and overcoming their memories. I will return to that important scene later.

5. claiming that a specific memory tells the truth, or

6. asserting that or asking whether a specific memory may be a lie. These acts are carried out most explicitly in two important scenes in the film. When Marius talks about a horrible dream dealing with boiling and eating the heads of people he killed during the war, he is asked whether this is a true story or a dream, and Patrick concludes this dialogue with the words "we want the real thing. We don't want a dream" (Julén/Niekerk 2010, 32:22). And after having discussed the role of his photographs of the war and his wish to burn these photographs with the three others, Marius says he is "feeling hurt about you guys not trusting me" (Julén/Niekerk 2010, 38:21).

7. claiming that something should be forgotten or left to oblivion. In *My Heart of Darkness*, this claim is not at first sight a political claim that demands oblivion from other groups or people. From the very beginning, Marius works upon his wish to get rid of his bad traumatic memories: "I don't want the memory any more" (Julén/Niekerk 2010, 34:00). He struggles with his memories in order to "reclaim my lost innocence" (Julén/Niekerk 2010, 3:10-4:20). I will discuss this claim for innocence later.

8. doing empirical memory research in the form of case studies. We can find even this act in *My Heart of Darkness*, because the film may be seen, especially by European spectators, as doing a kind of research in the field of memory studies by collecting information about the Angolan civil war, the events of these years, the different parties that fought against each other, the entanglement of Jonas Savimbi in international politics, and the role of international politics during the cold war.

9. acting in political struggles concerning national identity and current processes of reconciliation on national and transnational levels. This act can be found in the film especially when it is seen in the relative South African and Angolan contexts.[8]

Thus, in *My Heart of Darkness* we can identify many memory acts carried out as single acts, and furthermore we can identify other memory acts when considering the film on the whole. Of course we may treat the film as a representation 'of the war' or of the journey of the four veterans, but the point is that it is much more fruitful to consider it as a memory act or as carrying out several memory acts. As John Sundholm recently has shown in his interpretation of Peruvian filmmaker Cesar Galindo's most significant work, the five-minute *Fem minuter för Amerikas döda/Five Minutes for the Souls of America* (1992), a film can perform memory work

in order to hand over a narrative, an interpretation of the current state of a Latin-American country and to offer an experience to an audience of the dimensions of a colonial act. An act that has a past, present and a future, which the film depicts by using one continuing shot taken from a crane that moves slowly backwards. This film is only partly an object that circulates in cinemas and festivals it is also a

---

8   See Kaya de Wolff's paper in this volume; esp. the part about different readings of My Heart of Darkness.

conscious act of Galindo to organize and assemble a team to produce a story and an event, to intervene in Latin-American history and politics and to offer a story to an audience, even to Galindo himself. Not because the film represents a past, but because it enables a past and a present to unfold and to emerge in that present that takes place every time the film is being shown. (Sundholm 2016, 72)

In Galindo's film, it is the continuous movement of the camera and the film's use of images (Latin American, Christian, colonial) that can be read both as a way of tracing a story to its origin as well as revealing the hidden meaning of what is taking place on the screen, the colonial act and its consequences. In a parallel way, we can try to understand *My Heart of Darkness* not just as 'a film about the Angolan civil war', but also as a film that performs memory acts in order to hand over a narrative, an interpretation of what the Angolan civil war did to the veterans, and an experience offered to the audience in order to encourage processes of forgiveness and reconciliation. But compared to Galindo's work, at least in part, *My Heart of Darkness* falls back into presenting the wished result of the process of the veteran's journey. On the one hand, the process of the journey is shown to the spectator, but on the other hand, the intentions of Marius as the narrator, are quite clear from the beginning. This is why it is worth looking at some central contradictions that characterise the film.

CONTRADICTIONS AS A CENTRAL PART OF THE FILM'S AESTHETIC

My Heart of Darkness is a documentary about and entrapped in contradictions that I would like to exemplify by looking at a few of the key aspects of the film. A first contradiction or ambivalence concerns the aim of the film. In his article in this volume, Alexandre Dessingué demonstrates that the English version of the trailer clearly introduces the intention of the documentary in capital letters, "In search of reconciliation and forgiveness" (Julén/Niekerk 2010, 0'38"), while it can be asked whether and to what extent it really works as a process of reconciliation and forgiveness. This is what Kaya de Wolff (in her article in this volume) refers to when she notes the film's "inherent ambivalent polyphony".

On the level of the dialogues shown, there are of course many contradictions between the uttered opinions. Some of them are dissolved, and it can be asked whether the solutions take place only on the speech level, because the things shown on the visual level sometimes thwart them. One of the most controversial passages is the dialogue about the role of Marius' photos that starts after Marius' narration of his dream mentioned above (Julén/Niekerk 2010, 32:22). Patrick claims, "we want the real thing. We don't want a dream"; Marius then talks about his experiences as a member of the paratroopers during the war, and the fact that they did not know where they had been. And then, as if looking for a proof, he comes out with his photographs: "I brought pictures." Patrick replies "No, we don't need it". But Marius insists on showing the photos, opening his shoebox, weeping, and talking about his wish to get rid of his memories. And then he generalises his claim, switching from his wish about his personal memory to the memories of their group: "I want us all to get rid of our bad memories" (Julén/Niekerk 2010, 35:22). Patrick, Samuel and Mario express an important assumption about the photos and some doubt concerning Marius' position in the war: "He had a good life, being able to take pictures" (Julén/Niekerk 2010, 34:24); "If I go to war and I have a camera... That's not a war" (Julén/Niekerk 2010, 36:53). At the end of this scene, they come to the conclusion: "I believe you"; "I don't trust pictures". Thus the role of the photographs is very complex and ambivalent. Marius seems to use them as proof, the others do not trust the photographs; he wants to burn them, the others say "Don't burn it" and recommend he give them to his daughters. In this scene, no homogenous group is shown; there is Marius on one side and Mario, Samuel, and Patrick – the three Black men – on the other. Patrick makes it explicit when talking about "he" and "we three". In this respect, the colonial scene is repeated. Moreover, on the level of the aesthetic of the film as a whole, Marius' individual intention to get rid of the photos by burning them is foiled by the fact that he is one of the filmmakers documenting the photos, thus putting them in the scene and preserving them. Thus, rather than reveal the individual's expressed intention, the film proves its impossibility; Marius' wish for private ownership of his memories is a desire impossible to fulfill.

"A COMPLEX AND UNSTABLE PROCESS": RESIGNIFICATION AND REMEDIATION OF COLONIALIST AND RACIST IMAGERY

Remediations and resignifications are another important issue in *My Heart of Darkness*. Considering the series from Joseph Conrad's *Heart of Darkness* via Coppola's *Apocalypse Now* to Julén's/Van Niekerk's documentary, what are the results of the successive frameworks? The issues of colonialism and racism seem to be swallowed up by the "universal" aim of forgiveness; thus one could read a certain colour blindness into the work. Thus after some remarks about the theory of resignification and its foundations in the philosophy of language and discourse, this section will attend to the speaking positions of Marius and the other actors, the uses of daylight, firelight and darkness, and to the images of animals and nature connected with colonial imagery.

The onset of a theory of resignification can be found in Foucault's *The History of Sexuality: The Will to Knowledge* (Foucault 1978); Foucault himself does not use the term "resignification". The key passage is to be found in the context of Foucault's exploration of what he called the "repressive hypothesis" and his rejection of a conception of power as repressive (Foucault 1978, 81-86 and 92-101).[9] In contrast, Foucault here develops his concept of "power", by giving a list of very different elements that contribute to something like power in human societies rather than by giving a definition. He refuses to define an 'essence' or 'nature' of power, and thus he calls his theory a "nominalist" one: "it is the name that one attributes to a complex strategic situation in a particular society" (Foucault 1978, 93). Thus he talks preferably of "power relations". This leads to the question of resistance: Resistance should not be conceived as something external and opposed to power, "resistance is never in a position of exteriority in relation to power" (Foucault 1978, 95), but resistance does always take place from within power relations, although, at the same time, it shifts those relations. Foucault's explanation of the concept of resistance does not give essential features but rather an enumeration of different kinds of resistances:

---

9 Here I can only give a short description; a more detailed interpretation can be found in Hobuß, 2007 (in German).

there is a plurality of resistances, each of them a special case: resistances that are possible, necessary, improbable; others that are spontaneous, savage solitary, concerted, rampant, or violent; still others that are quick to compromise, interested or sacrificial; by definition, they can only exist in the strategic field of power relations. (Foucault 1978, 96)

Thus, resistance cannot be understood by an essentialist definition independent of situations, rather we have to think of resistances as plural, multiple and diverse practices. As a consequence, one of four rules one should follow when analysing discourses, is expressed by Foucault as the *"rule of the tactical polyvalence of discourses"* (Foucault 1978, 100) – it keeps us from unifying and normalising the variety of discursive elements in the relative contexts of different strategies. And here we find the key wording that expresses his aim:

It is this distribution that we must reconstruct, with the things said and those concealed [...]; with the variants and different effects – according to who is speaking, his position of power, the institutional context in which he happens to be situated – that it implies; and with the shifts and reutilizations of identical formulas for contrary objectives that it also includes. [...] We must make allowance for the complex and unstable process whereby discourse can be both an instrument and an effect of power, but also a hindrance, a stumbling-block, a point of resistance and a starting point for an opposing strategy. (Foucault 1978, 100f., emphasis added)

As I would like to show, such a "complex and unstable process" is to be seen at work in the case of the documentary *My Heart of Darkness*. In the text that directly follows the quotation, Foucault explains the discourse about homosexuality as an example for the discursive entanglement of instruments of power and resistance: The advance in the social control of homosexuality within the fields of psychiatry, jurisprudence and literature during the 19th century made possible "the formation of a 'reverse' discourse: homosexuality began to speak in its own behalf [...], often in the same vocabulary, using the same categories by which it was medically disqualified" (Foucault 1978, 101). In a condensed way, this passage details Foucault's idea of the investigation of discourses as a reconstruction of the varied ways and effects of discursive elements (as unfolded in his earlier works), as well as the new argument that such reconstructions can show how single "formulas", in the sense of definable discursive elements, undergo shifts that can – by repeating one and the same discursive element – serve contrary objectives. When speaking of the discourse as "both an instrument and an effect of power", single discursive elements and whole discourses can at the same time, and with identical instruments, exercise power as well as be understood as its results. In the case of homosexuality, first, power was exercised by certain discursive classifications in a discriminating way, second, "homosexuality" formed and constituted itself as a result of this discourse, finally, "homosexuality" used the categories invented by the discourse of discrimination and criminalization to speak about itself and claim rights, i.e. it established itself as "a point of resistance and a starting point for an opposing strategy". This understanding of counter-discourses is the starting point for a theory of resignification as elaborated systematically by Judith Butler.

In her book *Excitable Speech*, Butler studies the relationship between linguistic structures and social agency by investigating the injurious forces of discriminating utterances and hate speech. In Butler's thought, utterances of hate speech are the origin for possible resignifications. Reconsidering the utterances in Black rap music that uses the injurious word for Black people in order to gain empowerment and self-definition, she rejects two assumptions that cannot fully account for the resignification of the offensive utterance: On the one hand, the injurious force of some utterances could seem to depend completely on the context, and uses in new contexts could reinforce, minimise or cancel the effects. This is not fully the case as different people have different speaking positions. The other assumption claims that some particular utterances have their injurious forces *independent* of their contexts of usage; but this cannot account for the reinterpretation and re-connotation that sometimes take place. According to Butler, "the changeable power of such terms marks a kind of discursive performativity that is not a discrete series of speech acts, but a ritual chain of resignifications whose origin and end remain unfixed and unfixable" (Butler 1997, 14). Thus, interpellations, discrimination, and the injurious force of speech acts are not to be understood as sequences of *single* events of speech, but rather as a chain of elements connected to each other. Their connection is a matter of ritual repetition and iteration that in its performative aspect is characterised by simultaneity of repetition and new staging. But if the repetition is so important, one must ask how or when repetition of a linguistic expression is not merely a variation but can be qualified as critical, resisting or affirmative? There are no necessary conditions, but sufficient conditions can be given to make clear the state of the social frames and contexts of resignifying acts. There is one restriction: According to Butler, there is no such thing as a "neutral" or "innocent" repetition of injurious acts. The history of former uses always sticks, so to speak, traumatically to the repetition. Butler alludes to Freud: "there is no purifying language of its traumatic residue, and no way to work through trauma except through the arduous effort it takes to direct the course of its repetition" (Butler 1997, 38). From this perspective, every act of working through a trauma is at the same time its repetition, and the violation done by the injurious utterance can never be completely dissolved. The enterprise of resignification comes out as a "complex and unstable process" of subverting and opens up the political possibilities of appropriation but also involves certain risks and constraints.

So where are examples for acts of resignification in *My Heart of Darkness*? A first aspect is related to the chain of remediations of the central idea of a journey into darkness. One of the advertisements uses the following description:

In this extraordinary and revelatory documentary four war-veterans and former enemies of the South Africa/Angola conflict journey back to past battlefields deep within the African interior in search of reconciliation, forgiveness and possibly even atonement. A deep and characterful appraisal of what war does to men, it is a frightful insight into those who snap awake in the middle of the night, running, sweating, pursued ... terrified.[10]

It is the title of the documentary that first alludes to Joseph Conrad's novel and raises expectations that the journey in a boat up a river will be a central theme. This is enforced by the words "deep within the African interior". Seen in the light of theory of resignification as repetition, the trauma cannot be completely dissolved by the "journey to past battlefields". But it is not only the journey into past battlefield that the film offers; multiple montage sequences show animals and natural scenes that in the cultural memory are

---

10 https://www.journeyman.tv/film/5471/my-heart-of-darkness; last time consulted 1.9.2016.

connected with colonial imagery: landscapes, savannahs, trees against the backdrop of the sunset, birds, elephants and other wildlife. As Benedikt Jager shows in his article in this volume, "it seems to be impossible to evade the maelstrom of the exotic". "Journey" in this sense could even mean an enterprise like a tourist safari. But these pictures play several different roles: First, African nature is obviously the Other in oppositional montage to the war, the killings, the massacres and mass-graves. Second, some of the scenes imbue nature with healing power. Third, in some scenes, nature is not only idealized but shown as a mighty, dangerous power. For example, during and after the scene where Marius and Patrick are shown talking about their post traumatic stress disorder diagnoses in front of an archaic fireplace,[11] a herd of elephants is shown: in a first montage they drink at a watering hole, in a second, they leave the watering hole and move their large bodies with loud noises (Julén/Nierkerk 2010, 26:22). Here, the wild animals are shown as almost violent and mirror the aggression that comes up in the dialogue. Similarly, after the confrontation about the role of photography and the photographs I discussed above, there are two crocodile montages. Marius notices the crocodile at the turning point in the discussion.[12] The crocodile appears on the surface of the river, and the antagonism of the different opinions about trust, truth in memory, and reconciliation are unified; at the end Marius' position has won. In this case, nature metaphorically mirrors the process of the dialogue, or even seems to produce the movement.

One further example for acts or resignification is related to daylight, firelight and darkness that are used throughout the film in a very significant way. All the opening scenes in the first six minutes are set in full daylight: Marius showing the shoebox with his memories, the beginning of the journey in an open African savannah, Marius' introduction when his biographical information is given. Exactly when the three other veterans are introduced with the titles "Patrick", "Mario", and "Sammy", and just as Marius announces the journey as "a journey into the darkest of our hearts" (Julén/Niekerk 2010, 6:12), the light changes to the dark of the night, lit only by lamps. Until the turning point of the discussion introduced by the appearance of the crocodile (Julén/Niekerk 2010, 38:21), light and

---

11 See Benedikt Jager's paper in this volume.
12 See again Benedikt Jager's paper in this volume.

darkness are clearly distributed: Reports and confessions of the war and war crimes are spoken in the dark, mostly at the fireplace. The landscape and wildlife scenes are shown sometimes in full daylight, sometimes at sundown. After 37:46, the scene with the crocodile that marks the turning point of the discussion is shown in full daylight. This highlights the idea of *enlightenment* connected with Marius' idea of a combination of a Freudian talking cure and a banishment of bad memories. After this the film returns to framing scenes of talking at the fireplace or in the dark and scenes of landscape and wildlife in daylight or at sundown.

The concept of resignification oscillates between being a matter of linguistic structure or the materiality of the signs on the one hand and being the *social authorization* on the other hand. Austin always stressed the importance of social conventions as limiting and facilitating performative acts. In analyzing cultural memory, this can be useful in order to illustrate that there are limited speakers' rights, there are limited rights to perform certain memory acts. You need to be in an adequate position to perform certain memory acts. According to Jacques Derrida and Judith Butler, we can take into account further features of linguistic performativity in order to talk about other acts and practices as well, not only linguistic acts, something we must do to look at *My Heart of Darkness*. The most important feature is that contexts of uses cannot be fully controlled by single individuals in intentional ways. This is because linguistic meaning cannot be achieved by private acts of meaning, it exists only in forms of social practices. Thus, single individuals cannot fix meanings or new contexts arbitrarily. They are, however, responsible for their use of words, especially if the words and utterances have been used before in dangerous or controversial contexts, something Jacques Derrida treats with his concept of the "iterability" of signs (Derrida, 1988, 7-9). This concept implies repetition and difference at the same time, for reasons of the material structure of the act or utterance or performance that in each repetition is a new token of a recognizable and thus identifiable type. Because of this character, an utterance can be analysed in terms of the repetition it effects, i.e. its inevitable historical dimension, but also concerning its new appropriation of old concepts. In using signs and performing acts, we set them free, we spread them into the world, so that others can and will quote and repeat them again. Thus we cannot control or restrict the future understanding and uses of our signs, sentences, and actions. Speaking and acting are not conceived of as sovereign, autonomous practices that can be completely and intentionally controlled.

On the other hand, we can say that it is not the language itself that engages in these performative acts, but its users, a diference that highlights the importance of social agency and above all the role of social authorization as shown by Pierre Bourdieu:

> In fact, the use of language, the manner as much as the substance of discourse, depends on the social position of the speaker, which governs the access he can have to the language of the institution that is, to the official, orthodox and legitimate speech. (Bourdieu 1994, 109)

The characteristics of the material structure of meaningful items such as utterances, speech acts, and performances that uncover the importance of repetition are connected with the important role of social authorization. Speaker's positions are never distributed equally or non-hierarchical: Members of groups that are oppressed or discriminated against have other speaking positions and other possibilities of (re-)appropriation of concepts when compared with members of communities of perpetrators. This leads to a very complex situation concerning political utterances and (re-) appropriations. All this applies to cultural memory acts as well. Meaningful social practices like memory exist only in their relative contexts, stem from certain contexts, in some sense are repeated by us, and cannot be arbitrarily controlled. We can ask here not only about the rights of agents or speakers to do certain acts, but about the responsibility for these acts as well.

Because memory is based upon performative social practices, an individual or a group needs the social authority and must be in the right position to claim that certain memory acts should be done or what should not be remembered any more. Groups and agents who suffer violence, hate speech, or suppression are in a position to claim certain memory acts, while the perpetrators and related groups or agents are in a very different position. In the light of these thoughts, it is finally worth considering Marius' speaking position. On the one hand, he is one of the veterans that took part in the fights of the war: he killed other people and suffers from post traumatic stress disorder like the others. And during the journey, he always stresses the point that they need each other in order to reach forgiveness or at least freedom from trauma. But, as mentioned above,

Marius works upon his own wish to get rid of his bad traumatic memories – in order to "reclaim my lost innocence" (Julén/Niekerk 2010, 3:10-4:20). But "innocence" may be impossible to achieve. And the claim that the four veterans are a homogenous group suffers from a kind of color blindness characteristic of postcolonial perspectives. Marius is a White person, he is the initiator of the whole journey and the film production. Mario, Samuel and Patrick are Black persons. Marius describes his biography by referring to his (White South African) school, where "the brain fucking stated" (Julén/Niekerk 2010, 18:30) and his entry into the army, whereas Patrick and the others tell the spectators that they have been forced to join the fighting parties at the age of fourteen or fifteen and could not finish their education. To claim that they are equal is to neglect these differences. Though *My Heart of Darkness* can actually be read as a work of merit and a film about reconciliation and forgiveness, there is still some "secondary colonialism" (Kien Nghi Ha 2010) at work. But it is worth being studied as a narrative on its own in order to show how the memory acts are carried out: Rather than analyzing the film as a mere representation of certain memories or ways of coming to terms with them or as a representation of the historical truth – though these may be interesting questions –, the concept of memory acts allows to see its ambivalent and contradictory forces.

REFERENCES

Filmcredits

My Heart of Darkness – Sweden/Germany, 2010, 93 mins, directed by Staffan Julén and Marius van Niekerk. Coproduction of Gebrüder Beetz Filmproduktion, Eden Films, SVT and ZDF/arte. Supported by the Swedish Film Institute and the Angolan Ministry for Culture.

Literature

Austin, J. L. (1975). *How to Do Things With Words*. The William James Lectures delivered at Harvard University in 1955. Cambridge, MA and London: Harvard University Press.

Bourdieu, Pierre (1994). *Language and Symbolic Power*. Harvard University Press.

Butler, Judith (1997). *Excitable Speech: A Politics of the Performative*. London: Routledge.

Derrida, Jacques (1988). "Signature Event Context", in: J. Derrida, *Limited Inc*, Evanston, Illinois: Northwestern University Press, pp. 1-23.

Foucault, Michel (1978) [1976], *The History of Sexuality*, Volume 1: An Introduction. New York: Pantheon.

Hobuß, Steffi (2013), "Memory Acts: Memory without Representation.: Theoretical and Methodological Suggestions", in: *In Search of Transcultural Memory in Europe,*. Cost-Action ISTME IS 1203 Working Paper N. 1/2013. http://transculturalmemoryineurope.net/Publications/Memory-Acts-Memory-Without-Representation-Theoretical-and-Methodological.

Hobuß, Steffi (2013). "Repräsentationsarbeit als performative Praxis: Plädoyer für einen kritischen Repräsentationsbegriff", in S. Hobuß, P. Bell, K. Brandes et al. (eds.), *Die 'andere' Familie: Repräsentationskritische Analysen von der Frühen Neuzeit bis zur Gegenwart*. (S. 51-77). Frankfurt am Main: Peter Lang Verlag.

Hobuß, Steffi (2010). "German Memory Studies: The Philosophy of Memory from Wittgenstein and Warburg to Assmann, Welzer and Back Again". In A. Dessingué et al., *Flerstemte Minner* (pp. 22-34). Stavanger: Hertervig Academic.

Hobuß, Steffi (2007), "'Ein komplexes und wechselhaftes Spiel' – Sprachliche Resignifikation in Kanak Sprak und Aboriginal English", in: Russel West-Pavlov und Anja Schwarz (Hg.), *Polyculturalism and Discourse*, Amsterdam: Rodopi, S. 31-69.

Ka, Kien Nghi (2010), Unrein und vermischt. Postkoloniale Grenzgänge durch die Kulturgeschichte der Hybridität und der kolonialen "Rassenbastarde", Bielefeld: transcript Verlag.

Sundholm John (2016), "The memory practices of immigrant film-makers: Minor cinemas and the production of locality", in: *Crossings: Journal of Migration & Culture* Volume 7 Number 1, pp. 63-74.

Wittgenstein, Ludwig (2001). *Philosophical Investigations*. London: Blackwell Publishers.

Performing History

My Heart of Darkness from a Dramatist Perspective

KETIL KNUTSEN

In *My Heart of Darkness* pain is processed through documentary film. Along with three previous military opponents, Samuel Machado Amaru, Patrick Johannes, and Mario Mahonga, the South African veteran Marius van Niekerk returns to where they fought in the Angolan civil war of 1975 – 1988. He needs atonement. As a Swedish-German cooperation, the process is documented by Van Niekerk and Steffan Julen. From cinemas in 2010, it proceeded to online availability on the website Journeyman.

Angola has been relatively peaceful since 2002. A former Portuguese colony, it became independent in 1975. The independence led to a series of civil wars. The four veterans in the film represent the four sides of the first and major civil war from 1975 – 1991. Samuel Machado Amaru fought for the communist MPLA, Patrick Johannes for the anti – communist UNITA, Mario Mahonga on several sides and van Niekerk for apartheid South Africa. As the Cold War defined the period, the MPLA had Cuban and Soviet support while UNITA was backed by the US.

Painful memories of the war still influence the inhabitants, in particular veterans and their families. Since 2002, various reconciliatory attempts have been made. Understanding the dynamics between past, present, and future is crucial to understanding the role a painful past plays or might have played.

Traditionally, historical documentaries have presented viewers with results of historical research combined with the testimony of witnesses and historical reconstruction. By filming the the process the veterans went through dealing with their pain, *My Heart of Darkness* challenges viewers in new ways and explores new historical meaning and impact.

In order to understand the history didactics of the film, it is analyzed as historically created and historically creative through the dramatist analysis' focus on dominating driving forces. This means the film has a performative dynamic, where the ideas of past, present and future interact. First, I will discuss documentary film as history didactics and the historiographic context of *My Heart of Darkness*. Next, I will discuss pentadic analysis as dramatist interpretation of history didactics. The analysis itself is structured according to the five parts of the Burkian pentad: Scene, act, agent, agency, and purpose. Finally, I will discuss what motivates the film by clarifying what parts of the pentad dominate and what ratio (between the two most dominant parts) is most apparent.

THE HISTORY DIDACTICS OF DOCUMENTARY FILM

Documentary film demands a direct relationship to reality that we don't find in fiction. That engenders a trust in it along the lines of nonfictive literature. This trust, however, is somewhat misguided, according to the historian Robert A. Rosenstone, as a documentary film presents history in ways quite similar to fiction.[1] The past is actualized in different ways when various forms of media representation are employed.

However, my study is not based on Marshall McLuhan's view of media as the message, but on a more moderate take, where both the content and form of a historical account are influenced by the medium chosen.[2] As such, the perspective of the study is one of history didactics. The focus is not on the actual historical events or whether they are accounted for correctly as much as what representation strategies are employed.

Historical documentaries, TV series, and feature films have received criticism for reducing historical complexity and thus being inferior to written accounts. Public debate of this matter tends to revolve around

---

1 Robert A. Rosenstone, History on film. Film on history, (London and New York: 2006/2012), introduction.
2 Marshall McLuhan, Understanding Media: The Extensions of Man, (The MIT Press: 1964/1994), 7-21.

opinions on what is the most true account of a certain event. Some scholars, like the historian David J. Staley, revolve around logic, analysis, causality, events, diachrony, and linearity; visual history communicates analogy, synthesis, networks, thick deception, structure, the synchronic and nonlinear. According to Staley, that means visual accounts have a particular potential to show historical parallels, differences, change, and classifications, thereby allowing explorations of the multifaceted historical reality and challenges to traditional periodizations and chronology.[3] The film *My Heart of Darkness*, like other films, is a verbally and visually multimodal expression. Interpreting film is a matter of dealing with complexity.

In films, different media and modalities create meaning together. In her thesis on the history didactic potential of historical documentary film, Sara Brinch understands this dynamic as a correspondence between the horizontal and vertical meanings of the film, referencing Peter Larsen. Chains of meaningful expressions take their meaning from their juxtaposition to one another, as in the way a chain of images, music, or speech combine to constitute a certain communicative act.[4]

History and memory do not only enfold within a film, however. They also enfold socially, meeting various needs and functions. Not least within studies of memory and history use, a pronounced interest in the social characteristics of history takes place. The communication in films thus takes place on a societal level, as filmmakers employ a culture's available symbols, language and textual material. Simultaneously, film contributes to culture.[5]

---

3   David J. Staley, Computers, visualization, and history. How New Technology Will Transform Our Understanding of the Past (Armonk, London: M.E.Sharpe, 2003), 53-56.

4   Sara Brinch. "Historietimer for mediesamfunnet. En studie av En studie av dokumentarfjernsynets historieformidlende egenskaper" (Dr. art. avhandling, NTNU, 2003).

5   Bernard E. Jensen har written more about history as culture in , "Kampen om det historiedidaktiske historiebegreb," in Hvor går Historiedidaktikken? Kampen om det historiedidaktiske historiebegreb, eds. Sirkka Ahonen et.al. (Trondheim: NTNU. Sriftserie for historie og klassiske fag. No 45, 2004), 47-61.

My Heart of Darkness is particularly interesting as historical communication since it explicitly deals with how we handle history – the process of historization – and the history communicated as result and process simultaneously. The film is performative, in the words of Bill Nichols, who has categorized documentaries according to communicative mode. This means the film creator participates in the film as a constructor of a subjective historical truth significant to himself.[6] Thus, it may offer a unique perspective and present what historical knowledge can be.

The performative documentary is thus distinguished from an expository mode ("voice of god") where the audiovisual is inferior to a narrator driving the events: it is poetic with its exploration of the media's potential for expressing history as art; it is observant, the events presented as they happen without narration; and finally, it is reflexive, attention drawn primarily to the challenges of communicating historical events to recipients. These characteristics have dominated historical films more or less to this day. As a performative history documentary, *My Heart of Darkness* seems thus to represent something rather new.

Its novelty is apparent even in its title and trailer: The title alludes to an archetype or universal symbolism connected to something good being broken and in need of repair; the good being the heart and the damage being the darkness. The personal pronoun ("my") emphasizes the personal, which supports a performative view of history and invites expectations of a memorial, rather than historical, project. Such a view of history is also evident in the trailer, openly accessible on the Internet. Here, four veterans, former enemies, return to old war grounds seeking absolution and redemption, all there for the same reason – to embark and conquer the darkest of our hearts. The racist implications of the darkness as a frightening and destructive symbol echoes that of Joseph Conrad's colonialist novel, *Heart of Darkness*, and materializes as photographs in a shoebox and the boat's journey into the jungle.[7]

---

6 Bill Nichols, Representing Reality: Issues and Concepts in Documentary, (Bloomington and Indianapolis: Indiana University Press, 1991), 34.

7 Joseph Conrad, Heart of Darkness (Heritage Illustrated Publishing, 2012), Kindle edition.

HISTORIOGRAPHIC CONTEXT

Historiographically, *My Heart of Darkness* connects to the cultural turn of the 1970's, with its focus not so much on the past as on the meaning practices relating past to present and future. Also, it connects with the expressive manner of such practices and their place in history as a process. The term historical culture thus emphasizes our part in history as its creators as well as its creation. History becomes practice, something we do rather than something given.[8]

The film especially draws on schools of historical culture for thematic and theoretical angles: "history from below" and "oral history". The former regards acknowledging all voices, the latter, desegregating institutional practice (academia) and society. This type of historical account often draws on oral sources as certain groups lack written documentation. Thus, history from below and oral history are not merely research field and method, but also political, in documenting the memories of new groups, thus increasing their visibility. The sources are a combination of historian and interviewee.

Additionally, *My Heart of Darkness* connects to a traditional and still popular branch of the historical field; biography has long traditions back to prehistoric times. Biographic historization has always been about depicting an individual as part of a greater story. *My Heart of Darkness* departs from such traditions by emphasizing the use of history over the events themselves.

Thus, *My Heart of Darkness* connects to newer historical research on history use concerned with what needs history meets for individuals and groups as well as how various agents manage the past in order to meet such needs. At first glance, the film appears to be what Klas-Göran Karlsson calls a moral use of history, as it is a progressive attempt to recognize, revive, or rehabilitate history. Moral history use has particularly dominated various processes of reconciliation, such as public apologies, compensation, legal process, and truth commissions based on past transgressions.[9]

---

8   Jensen, "Kampen om det historiedidaktiske historiebegreb", 47-61.
9   Klas Göran Karlsson, "The Uses of History and the Third Wave of Europeanisation", in A European Memory? Contested Histories and Politics of Remembrance, eds. M. Pakier and B. Stråth (NewYork. Oxford: Berghahn

Coming to terms with a painful past constitutes a complex and diverse field of inquiry that has gained ground in the last decade. Research on reconciliation emphasizes critical perspectives on processes that allow communities to move forward. How to define reconciliation is still a matter of dispute. Cultural studies, psychology, communication, and memory studies have all been employed in studying how to constructively manage painful history.[10]

This field of research is anchored in the time after the second world war, with its trials. It is also anchored in the processes that took place after the fall of communism in Eastern Europe and the truth commissions in Bosnia, South Africa and Rwanda. These changes in how we relate to and research history represent a shift from summarizing to an interactive use of history.[11]

Until the 1970's, what dominated the presentation of history, in and outside the school systems, was accounts of political events and personages. To a certain extent history was equated with the past as part of an ambition to totalize history, as in the annales school. By turning attention to history use and individual memory, this study is founded on a history-didactic understanding of the term history, and thus represents a breach with traditional presentations of history.[12]

Books, 2010), 48-50. Timothy Garton Ash, "Trials, purges and history lessons: treating a difficult past in post-war Europe," in Memory & Power in Post-War Europe. Studies in the Presence of the Past, ed. J. W. Müller (Cambridge University Press, 2010), 265-283.

10 See Barbara Törnquist-Plewa and Eleonora Narvselius, "Cultural trauma theory and the memory of forced migrations. An example from Lviv," in Flerstemte minner, eds. Alexandre Dessingué et.al. (Stavanger: Hertervig Academic – Stavanger University Press, 2010), 35-54.

11 Jay Winter, "The Memory Boom in Contemporary Historical Studies", Raritan, Vol. 21, issue 1 (2001): 152-66, accessed August 13, 2015, URL: http://web.a.ebscohost.com/ehost/pdfviewer/pdfviewer?sid=3bfd7fc4-e748- 4c70-ae0b-f954bc30feca%40sessionmgr4002&vid=4&hid=4112

12 The didactic concept of history originates from Germany and Denmark, see Jörn Rüsen, Berättande och förnuft: historieteoretiska texter (Bokförlaget Daidalos, 2004), Bernard E. Jensen, Historie – livsverden og fag (København: Gyldendal, 2003), but stand in a larger epistemological context, see Paul Ricoeur, Time and

PENTAD ANALYSIS[13]

According to the dramatist theory of American philosopher and rhetoric scholar Kenneth Burke, life is a drama about human activity as involved in a conflict involving change anchored in language that has a conscious or subconscious purpose.[14] Burke defines language as symbolic action.[15]

A dramatist understanding of language (verbal and non verbal) sees it as a mode of action rather than a mode of knowledge.[16] This makes historical accounts symbolic action, meaning ways of life that explain situations and refer to corresponding action. To Burke, to name a situation is to simultaneously encompass it.[17] A filmmaker combines symbols in a documentary in order to understand and respond to certain problems, such as dealing with a painful past. Next, the film becomes a source of motivation for others.

From a dramatist perspective, historical account is as real as the events, rather than something attached to them. As performative, historical account may thus be understood as action connecting past, present, and future in a living dynamic.[18] As historical beings, we create history just as it creates

---

  narrative (The University of Chicago Press, 1984/1985), David Carr, Time, Narrative and History (Indianapolis: Indiana University Press, 1986).
13 Burkes dramatism is discussed in similar ways in Ketil Knutsen, "A history didactic experiment. Anno in a dramatist perspective", Rethinking History, Vol 20, Issue 3 (2016): 454-468, accessed June 13, 2016, doi.org/10. 1080/ 13642529.2016.1192258
14 J. R. Gusfield, Kenneth Burke: on symbols and society (Chicago: The University of Chicago Press, 1989), 17.
15 Kenneth Burke, Language as Symbolic Action. Essays on Life, Literature, and Method (Berkeley, Los Angeles, London: University of California Press, 1966), 63.
16 Kenneth Burke, A Grammar of Motives (New York: Prentice – Hall Inc.), XXII.
17 Kenneth Burke, The philosophy of literary form (Los Angeles: University of California Press, 1973), 109.
18 In context of the rise of postmodernity, researchers began to use the term performance in the 1970s, but the notion of performance can be traced back to Burkes dramatism. Partly as a result of the performative turn historians have studied rituals, gender, knowledge and memory, for more of "the performative

us. This can be connected to an ontological view of history as lived narrative. According to David Carr, who is a spokesman for such an understanding of history, history as narrative is told when it is lived and lived when it is told. It is impossible to live and then tell because life must have a narrative to be lived and narratives must have a life they are about. Life can then be described as a process in which we tell ourselves narratives.[19]

According to Burke, one may understand meaning and motivation for action by pentadic analysis.[20] The act – what happened? The scene – the temporal, spatial and social/cultural context of the act. The agent – the person primarily responsible for the act. Agency regards the means necessary for the agent to perform the act. The fifth part of the pentad is the purpose of the act.

While an internal-textual analysis explores domination in relation to one another, an external-textual analysis divides an act into the five parts of the pentad. The former is an interpretation of what Burke terms the motive of the symbolic act. This is done by interpreting what two pentadic aspects dominate the act and how they are connected. According to Burke, the main driving force of a symbolic act may thus be thoroughly explored. The aim of it is, according to Burke, to explain what is involved when we say what people are doing and why they are doing it. For instance, in a scene–act ratio, a scene causes an act. If the people of a country follow rules because the country is a comfortable place, this relationship may be argued by using this ratio. Contrarily, one may argue against an act-scene ratio if it is more likely that the country is comfortable to live in because the inhabitants follow the rules. Ratios are usually, but not always causalities.[21]

One has to interpret the most apparent act, agent, scene, purpose, and agency in order to find the main ratios in a symbolic act, such as a film. Though a difficult and confusing process, this distinction opens up inter-

---

turn in historical studies", see Peter Burke "Performing History: The Importance of Occasions," Rethinking History. Vol 9, No 1, Marsh 2005, doi DOI:10.1080/13642520420003229241, 35-52.

19 David Carr, Time, Narrative and History, 61-62.
20 The pentad is presented in Burke, A Grammar of Motives as the five key terms of dramatism.
21 Burke, A Grammar of Motives, 3 – 20.

pretation to a more dynamic treatment of relationships between memory and history, individual and collective history, result versus process, and sociality in relation to historicity. By examining the ways in which they may be interpreted in the dynamic between the five aspects of the pentad, the aim, rather than reaching a definite and final conclusion, is to expand our understanding of symbolic action, in this case as historical accounts. "What we want is not terms that avoid ambiguity, but terms that clearly reveal the strategic spots at which ambiguities necessarily arise."[22]

My interpretation of the film begins by discussing the complexity of each aspect of the pentad regarding *My Heart of Darkness*. Next, I examine domination in the dynamics between them, that is, what can be interpreted as the most emphasized aspects of the film in terms of pentad aspects. Examining each aspect continuality raises questions about the others as their dynamic is complex and close-knit. For instance, the scene of the act is past and present Angola. At the same time, present Angola serves as agency for the act, as van Niekerk travels there with the purpose of leaving behind his pain and regaining his lost dignity. Determining what two aspects dominate the entire film is done by looking at their overall presence combined with their connections to the other aspects. Various motives thus emerge. However, the task of interpreting which motive dominates presents the interpreter with lively resistance. Only after considering multiple alternatives may you achieve a pronounced impression of the main driving forces of the film.

On these grounds, the criticism of Burke feels familiar: "Everything implies everything else and everything is more complicated than it seems."[23] The challenge is that certain pentad aspects are so fundamental that they open up nearly unlimited possibilities for interpreting historical complexity. Digesting the complexity of each pentad aspect as independent of the film takes time, just as in the interpretative process. This study limits its scope to action as historically created, that is, action as a concern with time: past, present or future.

---

22 Ibid., XXIII.
23 William H. Rueckert, Kenneth Burke and the Drama of Human Relations. Second Edition (Berkeley, Los Angeles, London: University of California Press, 1963/1982), 267.

The following drama can thus be constructed out of *My Heart of Darkness* through pentadic interpretation: The dominant scene is the civil war in past Angola as expressed through the veterans', narrator's and visual narrative of the civil war events. The dominant agent is van Niekerk as a victim of war. This is expressed through his account of the pain he suffers in his deployment and how it has led to violence and a strong need for the kind of purgatory the film promises. This purge of pain from the past dominates as his act and is closely tied to finding something good that is lost. Memorabilia and archetypes are necessary for the act, but what dominates his agency is the archetypal voyage, not only a riverboat quest, but a journey in time. One may interpret it as a convergence of time and space. The purpose of van Niekerk's act is to reform his identity into one that fits a father. His war wounds thus cause his purgatory act of creating a film that converges time and space, with the end goal of fatherhood. Implicitly his act is necessary for his fatherhood.

SCENE: ANGOLA'S CIVIL WAR PAST

The scene of the film stretches across its temporal spatial backdrop from the 1975 Angolan civil war, where the characters fought, to their present voyage through Angolan villages and jungle. The film is thus anchored in the past and present as it rarely points to the future explicitly or to a change of past or present Angola.

Admittedly, a part of the scene includes Sweden, where van Niekerk fled, a place where he couldn't thrive. South Africa, where he grew up, is also mentioned as relevant in the formation of his identity and story. Additionally, other countries involved in the Angolan war figure. Angola, however, dominates the spatial aspect of the film's scene.

Present Angola, however, is emphasized more as a part of van Niekerk's agency than his scene. Present Angola gives his purgatory act a location to deposit his pain in exchange for a retrieval of his dignity. Past Angola dominates as scene as a backdrop for the act through the accounts of what happened there during the war, a painful, damaging history. This is expressed through animations presenting facts describing the sides of the war and their motives. Additionally, it takes form through the veterans and accounts of their experiences.

HISTORICAL AGENT: THE WAR VICTIM VAN NIEKERK

The veterans are depicted as victims of the war. From the start of the film, this tone is set as the following quote figures on a black background: "The line between good and evil is permeable and almost anyone can be induced to cross it when pressured by situational forces. Philip Zimbardo." Next, the veterans describe their loss, fear, and trauma and the destructive effects these have had on their relationships with others. The painful consequences of the war are audiovisually anchored in sequences of van Niekerk and the other three veterans with whom he travels. They are shown living apparently normal lives in a village. The audiovisual becomes anachronic. The juxtaposition of these visuals over the veterans' accounts gives an impression of simultaneity. They survived the war, but their prospects are damaged. More or less directly, this impression is reinforced by their comments.

Van Niekerk depicts himself as a victim in the dynamic of voice–over and authentic photographs from the past along with drawings. A multimodal presentation depicts photographs from his childhood and further life. As narrator, he describes a childhood influenced by the indoctrination of the apartheid regime in his native South Africa, where he was taught to view dark skin color as dangerous. This account is emphasized with sound; children's voices are replaced by more menacing sounds. The message appears evident: Van Niekerk's devolution from an intrinsically good, innocent child to a violent adult has environmental causes. As a whole, this ties the film's scenic aspect to Hannah Arendt's concept of the banality of evil, when evil is normalized through thoughtlessness and routine.[24]

Van Niekerk describes his intolerable situation in the present through an account of a violent incident: he hit and pushed his girlfriend in Sweden. He seems to attribute this incident to his feelings of isolation stemming from the memories of his painful experiences.

The ideas of the veterans as victims are strengthened through three authentic sequences from the war interwoven with van Niekerk's voice–

---

24 Hannah Arendt, *Eichmann in Jerusalem: A Report on the Banality of Evil* (Penguin Classics, 1994). Arendt's concept of the banality of evil is also discussed in Alexandre Dessingué's article.

over as narrator. The first sequence shows a mass grave next to some soldiers. Van Niekerk's voice–over says that war has always blurred the line between good and evil, creating monsters. He says that the mass grave included women and children, but took place before he was deployed. He points out that he disliked the way his commander took pride in the incident. The implication seems to be that the war and its commanders, rather than the individual soldier, are responsible for what happened.

The image of the soldier as victim continues in the second sequence from the war. We see UNITA commander Jonas Ziwimbi confidently and cheerfully surrounded by his soldiers. Van Niekerk's voice–over says that Ziwimbi continued the war through the ceasefire because the USA supported his cause as a crusade against communism. The war didn't end until he was killed.

That the painful past can still be felt is underlined by the third war sequence, showing parades of cheering crowds with flags. The narrator explains that people suffer even as politicians and generals in Angola have settled their differences. Many find themselves in the wrong place, hiding in exile or stepping on landmines, again, the visual combined with voice–over create an impression of simultaneity.[25] While experiencing joy over the war's end, people still feel its consequences. This way, the anachrony emphasizes the double challenge of war, nuancing the account of a war being destructive after its end.

Possible dominating agents in the film are the villagers or van Nieker's daughters, but their objectification rules them out as subjects with agency. The evil of the war is personified in its commanders, which makes them primarily attached to the scenic aspect of the pentad. The accounts of the veterans are complemented by the villagers as a generalized group in the present of the film. One exception is a wife who permits the veterans' journey. She more readily becomes a means of agency in the film, showing its message that the need for redemption is universal. Van Niekerk's daughters figure the most as part of the purpose aspect of the pentad, pointing to the future and his intentions toward establishing a sense of fatherhood to them through his process.

---

25 In her article, Nora Simonhjell discusses the representation of the disabled body and how this body was staged in a "beauty queen setting".

Thus, the veterans are most likely dominant agents of the film. Marius van Niekerk, Samuel Machado Amaru, Patrick Johannes and Mario Mahonga all act as participants in the purgatory process of removing pain and reclaiming dignity. Still, van Niekerk stands out from the group as more dominant than the rest. This is especially evident as he evades true dialogue with the others by arguing against their input and appears to prefers his own soliloquies. For example, he shows the others his photographs from the war in spite of their protests that the pictures lie. He is defensive, seeming uninterested in acknowledging their perspectives as options. The same critical attitude comes through when he points out their distrust in his lack of ability to recall exactly where he went during the war in Angola.

Van Niekerk's firm belief that all sides are alike, and all soldiers are alike in wanting forgiveness, projects his own need onto the other veterans. He assures the others of his intentions to secure this forgiveness in spite of their differing experiences and memories and their claims that the way he wants forgiveness is difficult for them. He stages a bonfire scene, where he not only throws pictures into the pyre, but demands the others do the same, in spite of their protests that he should hold on to the pictures. His memory is depicted as shared, and the pictures are symbols of this shared experience. The other veterans' collective role shifts towards the agency part of the pentad as they primarily become van Niekerk's rhetorical evidence of the universality of his experience. Rhetorically, he distinguishes them as useful only as parts of his victimized group.

Van Niekerk's domination strengthens as he controls the narrative as narrator and seems to structure its content to his own account. The other veterans are more pawns than contributors to the main narrative, which paradoxically needs to appear collective in order to seem a less subjective history. Thus, history from below in *My Heart of Darkness* paradoxically becomes history from above with van Niekerk's failure to relinquish his subjective superiority over his fellow veterans. Dividing lines from the past remain rather than dissolve, and this increases for van Niekerk's former enemies through the film. Van Niekerk's role as a writer and narrator as well as main witness of the past dominates. Eventually, Patrick Johannes gives in and says that all veterans do need forgiveness, what van Niekerk does is good and that he forgives him. It comes across more as polite sympathy than genuine collective experience, particularly as Johannes was the most critical member of van Niekerk's project.

ACT: PURGING THE PRESENT

Van Niekerk's perception of himself as victim drives his need for freedom from this feeling of victimization. This does not take the form of reconciliation as reconstructing the past with others in order to reckon with the pain and in order to improve the future.[26] Rather, it takes the form of a purgatory ritual and a destructive effort. Through acts that are archetypically known to destroy the problem, such as fire and blood, and through recovering something he lost through the journey, his main objective seems to damage the causes of his pain, these images and tense relationships he has with former enemies.

Burke often refers to drama as something ritualistic, which implies that the rhetor invites the audience to be unified in order to collectively participate in an experience. When Burke uses the term "ritual drama", it seems that he communicates actions that are explicitly symbolic and that have magical or religious functions for the participants, such as the Catholic mystery game or the ancient Greek tragedies. According to Burke, however, the ritual drama is a form of action that all human action originates from.[27]

---

26 Karlsson, "The Uses of History and the Third Wave of Europeanisation", 48-50.
27 Gretchen K. G. Underwood, "From Form to Function: In Defense of an Internal Use of the Pentad," K. B. Journal. The Journal of the Kenneth Burke Society Volume 7, Issue 2, Spring (2011): 1-4, accessed August 13, 2015, http://kbjournal.org/Underwood. Underwood draws heavily on Burke's The philosophy of literary form. Another treatment of ritual communication with reference to dramatism is Ingunn Kindervaag, «En retorisk redningsaksjon. Kongens minnetale etter 22/7, 2011,» Rhetorica Scandinavica nr. 64 (2013): 103-137. For more about the characteristics of history – based rituals, see Ketil Knutsen, «Historiebaserte politiske ritualer – hvordan politisk innflytelse utøves gjennom ritualer med historiske referanser», Memories we live by, eds. Janne S. Drangsholt et.al, (Hertervig Academic – Stavanger University Press, 2012), 11-23.

In the beginning of the film, van Niekerk's voice–over explains that unless he can find peace with his past, he will be an unfit father to his daughters. Over the images of a shoebox of memorabilia, including photographs of his daughters and the war, a knife and a badge, van Niekerk says that he wants his life and lost dignity back. Later, showing the war photographs to the other veterans, he says, "I want to show them and then get rid of them forever". The content of the shoebox are also shown at the bottom of shallow water while the narrator asks, "a knife, a medal, part of an ear, a finger, a trophy, a photograph. How is it possible to drag such baggage around for such a long time and not seeing it for what it is, it is to remind us to not forget."

Thus, the memorabilia serves as symbol of the burden that must be shed in the purgatory ritual. The memorabilia supplements the verbal, illustrating that van Niekerk's response to his painful past is a ritual, something performative that offers a release by way of representation. Handling a painful past becomes something manifestly visible and thereby manageable.

The purgatorial process happens by means of archetypes. One is the journey as a symbol of learning (moving from one place to another). This impression is strengthened by the narrator, who explains that a journey deep into the jungle is necessary in order to reclaim his innocence by returning the pain he got there. In this way, the dynamic between the verbal and the visual allow an understanding of a painful history through a universal pattern of action with long historical roots. The film is not so much about rational explanations or accounts as symbol-laden acts. In particular, this is illustrated by its end, where the veterans end up in a village where they perform a ritual purge by whipping themselves with goat blood while dancing.

AGENCY: CONVERGING TIME AND SPACE

The necessary means by which van Niekerk's gains agency for his purge (act) are the authentic sequences from the war and from Angolan villages, his use of archetypes and memorabilia, and his voice–over. The authentic war sequences, however, are primarily a part of the scene as they set the stage of the purge's backdrop archetypes while memorabilia constitutes a materialization of this purge.

The aforementioned means of agency are, however, inferior to the other veterans' enablement of van Niekerk's purge. Their supporting role is primarily instrumental in the film. Additionally, the film employs the means of shifting quickly between the journey into the jungle at night and in daylight. This emphasizes the narrators' recurrent focus on the tension between the pain and darkness of what is lost and the valuable, hopeful innocence of rebirth, illustrated by light.

Overall, the convergence of time and space describes the dominant agency of van Niekerk's act. This is because such an effect is necessary to construct the overall narrative. Angola's jungle represents the past, present and future. The past as the war scene, present as his journey's goal, and future as the place he feels he may reclaim his dignity. He portrays his liberation as dependent on a journey back in time. Such convergence of time and space create, according to Eviater Zerubavel, a direct material contact between present, past, and future at the same location, inserting the past into the present.[28] Angola works as part of the scenic part of the pentad, as mentioned, but more so as the place necessary for van Niekerk to converge place and time in his purgatory ritual.

PURPOSE: A FUTURE OF ENOUGH DIGNITY TO HEAL RELATIONSHIPS

The purpose of the film is dual: van Niekerk's wish to reclaim his pre-war identity combines with his aspirations of fatherhood. The former is consistently expressed through his voice–over's will to reclaim himself and move on to a better life in the future. This part of the film's communication works best distinct from the purgatory act of the present. The purpose is dominated by the future.

The second aspect of the purpose is expressed in three predominant ways: Initially, the voice–over of van Niekerk states that he has lost his family and stands to lose his daughters, unless he can come to terms with

---

28 E. Zerubavel, Time Maps. Collective History and the Social Shape of the Past (Chicago: University of Chicago Press), 41-43.

his pain and his past. This is illustrated by a photograph of his daughters in addition to the ones from the war and his other memorabilia. The contrast and connection between past and future communicate the war's ongoing impact on his close relationships.

Near the middle of the film, this is communicated to the other veterans, his purpose explicit. Towards the end, van Niekerk expresses hope that his daughters will be proud of him when they find out about his past and know that he has tried to repair it and let others do the same (referring to the other veterans).

An impression is made that he has in some ways lost some or all of his close relationships. Such a loss of influence adds an aspect of contact to the purpose of his purgatory ritual. His restraint on this topic may be either an attempt to refrain from giving away too much privacy or being selective on account of his difficulty entering the most dramatic core of his pain. Either way, an impression remains that relevant experiences are probably held back from his narrative.

The dual purpose of reclaiming his former self and being a father connect by the former being a prerequisite for the latter (to him). The former dominates the film quantitatively, repeated throughout, yet the latter overshadows it as providing a reason for wanting to reclaim his former self in the first place. That it is the other way around, makes fatherhood and contact with his daughters the dominating purpose of the film.

MOTIVE: HEALING ONESELF AFTER DAMAGE

In *My Heart of Darkness*, as in other films about society, the scene is always important. Although the scene – the war in Angola in the past – is important for the action, there are no ratios where the scene dominates. For example, there is not a scene–act ratio because although the war is important for van Niekerk's ritual purge, it functions only indirectly. The direct cause of the ritual is rather the agent – van Niekerk's, experience of being a war victim. Nor is it a scene–agent ratio. Admittedly, the film's presentation of the war's transformation of van Niekerk into a war victim is strong, but the scene is subordinate to the agent. A scene–agent dominance would have meant that that the scene influenced a transformation of van Niekerk. On the other hand, in the film the scene only becomes a frame of reference that van Niekerk does not take responsibility for. He relates to it passively, except as a means.

Neither is scene–agency striking in the film. It isn't the scene – the past war – which makes him create a film where he travels back to Angola and uses other people ritually, but rather his pain. Neither is the war the cause for his journey back to Angola to experience archetypical themes through his memorabilia, but the act – he undergoes ritual purification. Finally, van Niekerk's experience of having to do something extra in order to function as a father (purpose), originates not from the war, but from his own pain. Thus, it is not the scene that creates change in the film, but the agent and the act. The scene is not thematized in other ways than to show that van Niekerk is a war victim. As such, the function of the scene in the film is similar to the function of the agency.

The historical agent, van Niekerk, dominates the film mostly because all the other parts of the pentad are related to him. Agent–act – that he ritually purifies himself because he is a war victim – is most striking. Agent–scene is emphasized when van Niekerk interprets the war as something painful that hit him because he is in pain as a war victim. Thus, the war is interpreted in his image as something that has made him a victim. Agent–agency is emphasized as the other war veterans serve the function of rhetorical confirmation for van Niekerk's pain so that he is not carrying the responsibility for what happened alone. Thus, "history from below" becomes "history from above". The agent–purpose ratio communicates that in order to keep his self respect van Niekerk must function as a father for his daughters. However, this ratio is not strong since the relationship is not thematized in the film other than suggesting that van Niekerk is in pain because he can't be a father to his daughters.

The act, the purgatory ritual in the present, is like the agent, also strongly present in the film, but it does not relate as strongly to the other parts of the pentad as agent does. The act is mostly related to agency. The purgatory ritual causes van Niekerk to make a film where traveling back to Angola becomes a symbol of traveling back in time. If this ratio had dominated the film, it would have been more of an aesthetisation of history where the agency was more striking by coming into the foreground, for example if the film had been characterized by effects, which could have made it parodic, excessive, stereotyped, superficial, and more for show.[29] Since the act–agency ratio is so strongly present in the film it almost becomes thus.

The act–agent ratio, on the other hand, is weakly present because nothing in the film which indicates that the act – the purgatory ritual – has any impact on the agent, van Niekerk. The same applies to act–scene admittedly. The purgatory ritual does present the war as something painful, but it doesn't change the context. Van Niekerk is not interested in the scene other than disclaiming his own responsibility. Act–purpose is also present since van Niekerk undergoes a ritual purification because he wants to be a father again. However, this is only briefly mentioned and is not thematized in the film. If this ratio dominated, van Niekerk would come across as perhaps not using as many means, emphasizing the purpose.

The agency permeates the film since, as a medium and an audiovisual construction, it constitutes the agency. The agency–act ratio is expressed when van Niekerk makes the purgatory ritual possible by traveling back to Angola and employing archetypes and memorabilia. As such, the agency becomes the rhetorical means van Niekerk needs to express ritual purification as a process of change. The agency–scene ratio is expressed when the war is subjectively presented as an abuse of van Niekerk personally through authentic footage combined with voice–over and music. The agency–purpose ratio is expressed a few times when the voice–over says that van Niekerk wants to be a father for his daughters. There is no striking connection between agency and agent, except the mini-documentary in the film about van Niekerk's biography in South Africa, which presents him as a victim through the combination of "guilty" and "innocent" pictures and an ever faster and louder music as he gradually transforms from child to adult.

Archetypes and memorabilia and the journey back to Angola are selected as means because they are suitable for visual presentation. However, the instrumental function of the agency never becomes striking. The agency aspect is always weaker than the other parts of the pentad, even if the means are strongly expressed in relation to the act, the purgatory

---

29 Jörn Rüsen, Berättande och förnuft, 160-172 discusses the meaning construction of historical consciousness in three dimensions: The esthetic, the political and the cognitive.

ritual. If the agency had dominated, the medium, in the words of Marshal McLuhan, would have been the message of the film too a much greater extent.[30] Additionally, the film would have been to a much greater extent an aesthetic experience.

Like the scene, so also the purpose: Van Niekerk's desire to be a father for his daughters in the future plays a subordinate role in the film, but has an even weaker presence than the agency. The purpose–agent, that van Niekerk's desire to be a father for his daughters reflects his pain, is admittedly clearly defined in the film, but not striking because it is not thematized much and the purpose seems not to influence van Niekerk more than one could reasonably expect. His desire to be a father for his daughters is not connected to the scene, or the war in the past, or any other temporal or spatial context. If van Niekerk had communicated a will to change the context, the scene would have been more characterized by the purpose, for example, his daughters' childhood. However, when it comes to purpose–act, it is communicated that van Niekerk undergoes a ritual purification because he wants to be a father for his daughters, but this is not a major theme and nothing in the film suggests that it works. Nor does anything suggest that the agency specifically makes it possible for van Niekerk to be a father for his daughters. Generally, effects or results of the action are outside the film. The film only alludes to it and the viewers learn little about it. In a film so enclosed within itself as *My Heart of Darkness*, the purpose can't be as dominating as for instance in a documentary by Michael Moore, where political change is important.

It is the historical agent, van Niekerk, and secondly the act, the purgatory ritual, that define the film. Agent–act dominates because the causality between these are the strongest. The film is more about what happens to whom during the process than the scene, purpose, or means of agency for it. The agent dominates the most because it connects the scene and the act. Act is second because the purgatory process in itself is central. As agent, van Niekerk purges himself because of his personal quality of victimhood. In my impression, the film is more about managing damage than taking responsibility or blame for it.[31] Van Niekerk primarily takes

---

30 McLuhan, Understanding Media, 7-21.
31 Also, Benedikt Jager and Alexandre Dessingué argue in their articles that the film is not about forgiving as it proclaims.

ownership of the repair process. Thus, the anticipating and remembering function of history are placed in the background of its diagnostic function, since the film is more about dealing with the present than interpreting the past or expecting something from the future.[32]

Conclusion

By pentadically interpreting the film as history created and history creative, this study deepens and nuances our understanding of a difficult past. It does this by exploring the complex interplay between present, past, and future on many levels. A war veteran who feels damaged by a painful past wants to eliminate the causes of his pain and recover his dignity. He does this by converging time and space, his veteran stories a proof of his own painful history and that the responsibility is not his alone. His act points towards the future in his desire to be a father to his kids. Thus, the relationships between past, present and future, individual and collective history, memory and history, interpretation and selection, result and process, and sociality and historicity are discussed.

The film reconstructs the story not through reflection, but deconstructs it by accounting a painful past. It is thus not what Klas-Göran Karlsson calls moral use of history, which is about drawing lessons from the past in order to construct a better future.[33] The film's intent as an individual project and its rejection of historical objectivity make it more a testimony of a memory than history. The film can also be defined as amoral through van Neikerk's dominant role and his use of other people (weaker parties of the war). At the same time van Niekerk is also a victim, both as a soldier and while growing up and living in South Africa's oppressive apartheid regime.

The film may remind us of non–use of history or silence in that van Niekerk refuses to remember what happened historically, rejecting it to argue that he is a victim, so only his individual memories are crucial. He also says directly that it is not history, but memories that matter. Thus, the process of purification is neither about forgetting nor remembering, but a form of silence. Jay Winter defines silence as the space between

---

32 Jensen, Historie – livsverden og fag, 68-69.
33 Karlsson, "The Uses of History and the Third Wave of Europeanisation", 48-50.

remembering and forgetting.[34] Van Niekerk cannot remember the good without forgetting the bad. Thus, the film is set in the tension between memory and forgetting.

This may not only be associated with postmodernism and the cultural and linguistic turns of the humanities, but also modernism. The ritual staging involves an essentialism. To deal with a difficult past is to find something basic – an essence that already exists in some underlying reality. The film stages views of change and history as parts of something underlying and static – all just shadows of the essence.

Essentialism has historical roots back to the nation's construction in the nineteenth century. Community was not something you chose but were born into, as in the difference between German nationality and French nationality. Historically, this way of thinking has been used to assimilate or exclude those who were different.[35] At the same time the film may be associated with modern historical writing, history from below, though van Niekerk's dominance becomes history from above.

The film's essentialist dimension also gives it an ahistoric dimension since the claim is that there exists a truth that is time–independent. Central in historical thinking is the time dependence of all phenomena, their particularity. That does not mean that there may not be similarities and trends over time, but that what happens always must be understood in the light of the context of time and space.

This study's insights have transfer value to other areas of coming to terms with a difficult past, whether it is educational, political, memorial, or genealogical investigation. Rewriting Bernard Eric Jensen's term, contemporary history culture is revisionist, constantly rewritten and renegotiated.[36] Pentad interpretations of history didactics may improve

---

34 Jay Winter, "Thinking about silence," in Shadows of war: a social history of silence in the twentieth century, eds. Efrat Ben-Ze'ev et.al. (Cambridge: Cambridge University Press, 2010), 4.

35 The role of history in democratic processes is discussed in Bernard E. Jensen, «Medborgerskab historisered», in Skolen, nasjonen og medborgaren, eds. Trond Solhaug et.al. (Trondheim: Tapir akademisk forlag, 2012), 99-110. Jensen discusses essentialist history as opposed to constructivist history, which he believes is more consistent with living in a multicultural and globalized society.

36 Jensen, Medborgerskab historisered, 105-108.

manageability of the various and complex situations and give insights into performative history.

REFERENCES

Arendt, Hannah. *Eichmann in Jerusalem: A Report on the Banality of Evil.* London: Penguin Classics, 1994.

Ash, Timothy G. "Trials, purges and history lessons: treating a difficult past in post – war Europe." In *Memory & Power in Post – War Europe. Studies in the Presence of the Past*, edited by J. W. Müller, 265-283. Cambridge: Cambridge University Press, 2010.

Brinch, Sara. "Historietimer for mediesamfunnet. En studie av En studie av dokumentarfjernsynets historieformidlende egenskaper." Dr. art. avhandling, Norges teknisk – naturvitenskapelige universitet, Trondheim, 2003.

Burke, Kenneth. *A Grammar of Motives.* New York: Prentice – Hall Inc, 1945.

Burke, Kenneth. *Language as Symbolic Action. Essays on Life, Literature, and Method.* Berkeley, Los Angeles, London: University of California Press, 1966.

Burke, Kenneth. *The philosophy of literary form.* Los Angeles: University of California Press, 1973.

Burke, Peter. "Performing History: The Importance of Occasions." *Rethinking History.* Vol 9, No 1 (2005): 35-52. Accessed June 13, 2016. doi 10.1080/1364252042000329241, 35-52, 2005.

Carr, David. *Time, Narrative and History.* Indianapolis: Indiana University Press, 1986/1999.

Conrad, Joseph. *Heart of Darkness.* Heritage Illustrated Publishing, 2012. Kindle edition.

Gusfield, J. R. *Kenneth Burke: on symbols and society.* Chicago: The University of Chicago Press, 1989.

Jensen, Bernard Eric. "Kampen om det historiedidaktiske historiebegreb." In *Hvor går Historiedidaktikken?* (No 45 Skriftserie for historie og klassiske fag), edited by Sirkka Ahonen, M. Poulsen, O. S. Stugu, M. Thorkelsson & U. Zander, 47-63. Trondheim, 2004.

Jensen, Bernard Eric. "Medborgerskab historisered". In *Skolen, nasjonen og medborgaren*, edited by Trond Solhaug et.al, 99-110. Trondheim: Tapir akademisk forlag, 2012.

Jensen, Bernard Eric. *Historie – livsverden og fag*. København: Gyldendal, 2003.

Karlsson, K.-G. "The Uses of History and the Third Wave of Europeanisation" In *A European Memory? Contested Histories and Politics of Remembrance*, edited by M. Pakier and B. Stråth, 38-55. NewYork, Oxford: Berghahn Books, 2010.

Kindervaag, Ingunn. "En retorisk redningsaksjon artikkel. Kongens minnetale etter 22/7, 2011." *Rhetorica Scandinavica* 64 (2013): 22-40.

Knutsen, Ketil. "Historiebaserte politiske ritualer – hvordan politisk innflytelse utøves gjennom ritualer med historiske referanser", in *Memories we live by*, edited by Janne Stigen Drangsholt, Benedikt Jager og Anne Kalvig. Hertervig Academic – Stavanger University Press: 11-23, 2012.

Knutsen, Ketil. "Strategic silence – Political Persuasion between the remembered and the forgotten." In *Beyond Memory. Silence and the Aesthetics of Remembrance*. Series: Routledge Approaches to History, edited by Alexandre Dessingué and Jay M. Winter, 125-141. Routledge, 2015.

Knutsen, Ketil. "A history didactic experiment: the TV series Anno in a dramatist perspective." *Rethinking History*. Vol 20, Issue 3 (2016): 454-468. Accessed June 13, 2016.

Larsen, Peter. *TV-analyse for mediestuderende. Tre kapitler om betydning og fortælleanalyse*. Bergen: Universitetet i Bergen, 1995.

McLuhan, Marshall. *Understanding Media: The Extensions of Man*. Cambridge, London: The MIT Press, 1964/1994.

Nichols, Bill. *Representing Reality. Issues and Concepts in Documentary*. Bloomington and Indianapolis: Indiana University Press, 1991.

Rosenstone, Robert. *A History on Film. Film on History*. Series: History: Concepts, Theories and Practice. 2nd Edition. London and New York: Routledge, 2012.

Rüsen, Jörn. *Berättande och förnuft: historieteoretiska texter*. Göteborg: Bokförlaget Daidalos, 2004.

Ricoeur, Paul. *Time and narrative*. Chicago: The University of Chicago Press, 1984.

Rueckert, William H. *Kenneth Burke and the Drama of Human Relations.* Second Edition. Berkeley, Los Angeles, London: University of California Press, 1963/1982.

Staley, David J. *Computers, Visualization, and History: How New Technology Will Transform Our Understanding of the Past.* Series: History, Humanities, and New Technology. New York: M.E. Sharpe, 2003.

Törnquist-Plewa, Barbara and Narvselius, Elonora. "Cultural trauma theory and the memory of forced migrations. An example from Lviv." In *Flerstemte minner,* edited by Alexandre Dessingué, Ann Elisabeth Laksfoss Hansen and Ketil Knutsen, 35-54. Stavanger: Hertervig Academic – Stavanger University Press, 2010.

Underwood, Gretchen K. G. "From Form to Function: In Defense of an Internal Use of the Pentad." *KB Journal,* Volume 7, Issue 2 (2011). Accessed October 10. 2016. URL http://kbjournal.org/Underwood.

Winter, Jay. "The Memory Boom in Contemporary Historical Studies." *Raritan,* Vol. 21, issue 1 (2001): 152-66. Accessed August 13, 2015. URL: http://web.a.ebscohost.com/ehost/pdfviewer/pdfviewer?sid=3bfd 7fc4 – e748 – 4c70 – ae0b – f954bc30feca%40sessionmgr4002&vid= 4&hid=4112.

Winter, Jay. "Thinking about silence." In *Shadows of war: a social history of silence in the twentieth century,* edited by Efrat Ben-Ze'ev, Ruth Ginio and Jay Winter, 3-32. Cambridge: Cambridge University Press, 2010.

Zerubavel, E. *Time Maps. Collective History and the Social Shape of the Past.* Chicago: University of Chicago Press, 20.

The Role of Music in Memorial Production and Discourse in *My Heart of Darkness*

DAVID-ALEXANDRE WAGNER, JON SKARPEID

FILM AND MEMORY STUDIES

The role of film in shaping individual and social forms of memory is widely acknowledged by researchers within the field of Memory Studies. Astrid Erll (2010, 2011) has presented the notion of memory as mediated and categorized the different modes of representation and the mediated nature of cultural memory in literature and film. Susannah Radstone (2010) has described and underlined the complex and deep interpenetrating ties between cinema and memory. Moreover, she has stated that what is at stake is the question of memory's "transindividuality" through cinema (and other media): the relationship of the personal and individual aspects of memory with public and collective memory.

A central and productive reference on this matter is Allison Landsberg's notion of prosthetic memories (2004). Landsberg has argued that a specific form of memory emerges from the sensual experience about the past a film may provide a spectator. As an embodied experience, watching a film enables the audience to identify with events and characters represented on the screen. This can create vivid memories, even from events the spectator did not personally lived.

For Radstone, Landsberg's theory implies little difference between a cinematic memory and a personally experienced memory. This results in both overestimating cinema's powers and underestimating the interplay between text and spectator. Radstone prefers the more complex exchanges between cinematic images and personal memory described in Anette Kuhn's (2002) and Victor Burgin's (2004) cinema/memory theories, i.e. film images as "screen memories" that become integrated with the other memories of our own individual inner worlds. For us, this is where the traditional field of memory studies meets cognitive sciences (as emphasized by Ib Bondebjerg (2014, 13-14)). We will argue that it is irrelevant to establish a frontier between mediated memories and personally experienced memories, not only because this frontier has proven to be thinner under the challenge of growing virtual reality technologies, but also merely because this difference is not really relevant at a cognitive level. Every memory, on an individual basis, is the product of an experience, whether it is mediated or remembered as personally experienced. What seems crucial for the *production* of individual memories is not their origin or their nature *per se*, but the emotional charge and/or the anticipated future value of the experience that produced them (Desgranges and Eustache 2010, 453-454). Whether an experience, an event or a fact will be selected and stored as a memory will depend on the emotional load it has carried or the utilitarian value it is expected to retain in the future. Later, a memory will be all the more pronounced and vivid if it is recalled and nurtured, and doing so will noticeably modify and alter it. The recollection of old information happens in interaction with the present context and other memories, which leads to a reconstructed, reinvented, and distorted remembering (Markowitsch 2010, 280-282). That is where the power of the cinematic experience is fascinating: it can convey through its forms and narratives an emotional load that is in line with, and we would say sometimes even stronger than, "real life" experiences. Moreover, considering that narratives are one of the central means by which we, as human beings, make sense of what we experience, films play a crucial role in activating and forming our memories through the emotions their narrative (and rhetorical) structures trigger. Memory is then also a narrative and emotional activity (Bondebjerg 2014).

WHY IS *MY HEART OF DARKNESS* AN INTERESTING CASE FOR MEMORY STUDIES?

To illustrate and explore the relationships of memory and cinema, *My Heart of Darkness* is particularly interesting through its potential to produce prosthetic memories and through its reflexive discourse on memory. The director/narrator/main character of the film, Marius van Niekerk, is a former paratrooper of the South African apartheid regime, exiled to Sweden for political reasons, and suffering of a post-traumatic stress disorder since his implication in the Civil War in Angola. This documentary is about his quest for forgiveness. He returns to Angola to organize a boat trip upon a river along with three black veterans from both sides, a trip in which they discuss and exchange testimonies about their war experiences. The journey functions as a quest for mutual understanding, relief, and reconciliation on both a material and symbolic level. It reaches its acme towards the end with a cleansing ritual meant to complete the healing process. As an autobiographic film – narrated at the first person, but also polyphonic, through the testimonies of the other protagonists – *My Heart of Darkness* is a memory-reflexive documentary about a soldier's relationship to traumatic war memories. However, it is also a memory-productive film, through its obvious remediation of central literary and cultural masterpieces of the Western world's collective memory. Its very title, *My Heart of Darkness* refers to Joseph Conrad's renowned novella. It also inevitably invokes Francis Ford Coppola's *Apocalypse Now* (1979) in many ways. The opening scene of *My Heart of Darkness*, set in the main character's hotel room, framed by sounds of war and the voice-over of the narrator, the initiatic character of the following boat-trip along the river during the whole film, the use of nature as a metaphor, and the final ritual ceremony of sacrifice/purification are all clear parallels to *Apocalypse Now*. Moreover, the main character Marius can be connected to Capt. Willard, played by Martin Sheen, in his mission to find and kill Colonel Kurtz, a mission symbolizing the darker side of human nature when confronted with extreme situations. For Marius van Niekerk, this "killing" will be finding a way to deal with his traumatic memories. These interfilmic and intertextual references, along with the whole network of their premediations and remediations, strengthens the message of the film and its scope, binding together the traumas of the less known Civil War in Angola with those of the conflict in Vietnam and, more generally, with the global issue of colonization.

FILM MUSIC AND MEMORY STUDIES

Nonetheless, a particular element in the narrative structures used in *My Heart of Darkness* has especially drawn our attention: the role of music in relation to the memory-productive and memory-reflexive functions of the film. This article has a double intent: first, to heighten the decisive impact music may have as one of the strongest sources of emotion on the memorial production and discourse of a film; and second, to exemplify and analyze this impact in the case of *My Heart of Darkness*.

Music activates emotions or moods – sadness, anxiety, nostalgia or melancholy, suspense, joy, happiness – that strongly influence the meaning the listeners assign music (Juslin and Sloboda 2001, Meyer 1956). Music appears to mimic some of the connotative features of language and to convey some of the same emotions that vocal communication does but in a non-referential and non-specific way. It also involves some of the same neural regions that language does; however, far more than language, music taps into primitive brain structures involved with motivation, reward and emotion (Levitin 2011, 185-187, Juslin and Västfjäll 2008). Consequently, the question of whether music really induces emotions is still warmly debated, but seems also more and more obvious (Juslin and Västfjäll 2008, 561-562, Hunter and Schellenberg 2010), especially in relation to visual media (Cohen 2001).

Although there seems to be a lack of writing on the varied functions of music within documentaries (Corner 2005, 242), film music in general is not a neglected art anymore, and music's place within cinema is now frequently analyzed by academics interested in cinema and film writers (Dickinson 2003, 1). The influence of the soundtrack is widely acknowledged in film. The function of film music can be very diverse, and as such, has been subject to various studies and categorizations by scholars.[1] However, we will refer to Annabel J. Cohen's (1999) list of eight functions in a cognitive perspective, as the clearest and the most exhaustive we found: 1) Music masks extraneous noises; 2) Music provides continuity (or discontinuity) between shots; 3) Music can attract direct attention to important features on the screen; 4) Film music induces emotions, moods and feelings; 5) Film music conveys meaning and furthers the narrative, especially in ambiguous scenes; 6) Through association in memory, music becomes integrated with the film and enables symbolization of past and future events; 7) Music heightens the sense of reality of the film or the spectator's absorption into the film; 8) Music adds to the aesthetic effect of the film.

All these functions, except the first one, are relevant in the context of memory studies, because they involve memory processes or are involved in the activation of the film's potential as producing prosthetic memories. However, functions 5) and 6) are particularly important because the spectators have a tendency to experience and interpret the presence of music and images as bearing a meaning, probably because we are fundamentally meaning-seeking as human-beings, as a result of evolutionary processes. We expect music to have a meaning and an intention in film. We have a tendency to pair music and images and to build a meaningful analogy between their structural relations. As a result, we aim at unveiling/interpreting the analogic or metaphoric relation between music and the images/the narrative it is attached to (Larsen 2013, 225-226). As Annabel Cohen (2001, 254) emphasizes:

[...] the analysis of the acoustical information must be regarded as a preattentive step that leads the listener to inferences consistent with the diegetic world of the film. From moment to moment, the audience member extracts information from non-diegetic sources to generate the emotional information he or she needs to make a coherent story in the diegesis.

Function 7) is also central, although it is often forgotten or neglected. By contributing to the abolition of the frontier between the film's reality and

---

1 Among them, we can mention Kurt London, Zofia Lissa, Claudia Gorbman, Katryn Kalinak, Michel Chion, Johnny Wingstedt, Peter Larsen and Annabel J. Cohen.

the spectator's reality, film music plays a crucial role that often seems specific for feature films.[2] Indeed, the implicit knowledge about a feature film is that it is not for real; it is actors playing a part and special effects and stunts carefully realized by professionals. Music helps temporarily erase the consciousness that what we see is actually a movie. In fact, music is such a characteristic of the fictive film that it is usually absent from documentary films, or fictive films with an intention of being closely wired to reality. For a spectator who is ignorant of the type of film he is about to watch, the music score of the opening images can function as a clear signal that the film is fictional. As a consequence, many feature films wanting to look like documentaries will avoid using music, and when music intervenes in a documentary, this gives the impression of a fictionalized scene. Film music can even be so powerful that it may produce a feeling "larger than life" (Gorbman 1987, 68, Larsen 2013, 174). Actually, this feeling may temporarily go beyond the perception of a lived reality, shaping a more vivid feeling of experience than what real life seems to offer us: it produces a feeling of augmented (virtual) reality. When strongly associated with situations or characters in a movie, it can really shape important prosthetic memories. In many cases of fandom, the movies are so frequently viewed, commented upon, and remediated that they constitute an important part of a person's set of real-world references. These effects of music in series like *Star Wars*, *Indiana Jones*, *The Hobbit*, or *The Lord of the Rings* are clear examples of this. Most of us are aware that these are not "really lived" memories, but they still play an important role as set of references in the cultural memory of many individuals, partly enhanced by their music scores. Even if you mime an adventurer's peripeteia or a sword combat between two knights, the scene will not be recognized as rapidly as if you hum the theme of *Star Wars* or *Raiders of the Lost Ark*.

Many studies have demonstrated the impact of music on the interpretation of visual content and on memory (Hoeckner and Nusbaum 2013, Bezdek and Gerrig 2008, Cohen 2001). Music has not only the ability to influence the interpretation of a visual presentation (Marshall and Cohen 1988), it heightens the emotional impact of a mood-congruent scene, and attenuates the emotional impact of an overtly incongruent scene (Bolivar, Cohen, and Fentress 1994, Baumgartner et al. 2006). Moreover,

---

2 Perhaps by augmenting arousal and by captivating attention to the scene.

by promoting certain inferences, film music serves to provide an interpretative framework to neutral or ambiguous scenes, or to interpret scenes to come (Bullerjahn and Güldenring 1994, Vitouch 2001, Tan, Spackman, and Bezdek 2007). In addition, film music has an influence on how well a scene is remembered (Boltz, Schulkind, and Kantra 1991, Boltz 2001, 2004). When music accompanies a scene, the scene is better remembered if the music is mood-corresponding.³ Film music has thus an obvious impact, amplifying or undermining a film's potential to produce memories.

THEORETICAL FRAMEWORK

Cohen's Congruence-Association Model

Annabel J. Cohen (2013) has reframed the role of music within film in a congruence-association model where inferences are made on a structural and semantic basis and help the spectator to establish a working narrative. She identifies *internal semantics* as what the spectator feels while watching a film, and *external semantics* as the meanings, rules, and cultural conventions learned from experience.

Internal semantics is rooted in physiological processes. Structural features are linked to emotions on a visceral and almost automatic level. Cohen (2013), referring partly to Gorbman's *Unheard Melodies* (1987), underlined that music doesn't need to be consciously listened to, to have an effect. For example, discordant notes will provoke a feeling of

---

3 However, Boltz did not test the effect of mood-divergent music in extreme contrasts, like the "anempathic" or "ironic contrast" present in some cases (Bordwell and Thompson 1979, Gorbman 1987). A classical example of this effect is the murder scene in Clockwork Orange (1971), accompanied by Rossini's The Thieving Magpie, where the contrast between the playfulness of the soundtrack and the violence of the scene is patent. Music gives a burlesque tone to the scene that is in line with the clown-like outfit of the intruder and conveys a more subtle meaning. The severe contrast makes the scene very memorable in its surrealistic weirdness, even though the incongruence of the music still attenuates the emotional load of the sequence.

unpleasantness, and a succession of notes in low pitch at a higher tempo than the "standard" human heartbeat (ca. 60) will elicit a sense of danger, uncertainty, and anxiety. Boltz (2004) considers these musical characteristics as universal invariants, similar to the interpretation of gaits (Montepare, Goldstein, and Clausen 1987) or of facial expressions (Ekman and Friesen 1975, Ekman 2007). It is fascinating to assume that the effects of these fundamental musical characteristics really are universal invariants because it implies that some parts of the memorial processes elicited by these emotions are universal.

External semantics is rooted in cultural codes and conventions. It could be the pairing of a theme with a character or an idea, like in a leitmotiv, or cultural conventions associated with certain music styles. A classic example is the use of Edvard Grieg's "In the Hall of the Mountain King" as a leitmotiv associated with the crimes that will be committed by the main character in Fritz Lang's M (1931). Mendelssohn's "Wedding March" from a *Midsummer Night's Dream* (1826) is also often used to evoke marriage and a happy ending.

A major characteristic of film music is to bridge these two dimensions. In addition, these two dimensions function at two different levels: a congruent and an associative level. *Congruence* is a product of structural parallels between music and picture, i.e. the structural features of the soundtrack match the structural features of the images. Cohen's hypothesis is that this matching triggers similar neural excitation patterns of auditory and visual cell networks in the brain:

Assuming that watching an action elicits the same neural activity as enacting the action oneself, the engagement of the mirror neuron system is likely to be a major contribution to the internal semantics of the film experience. [...] Because the direction of the melody can also engage the same mirror neuron system (Overy and Molnar-Szakacs 2009), film music can reinforce the empathetic effect. [...] Once attention is directed by visual information, which shares structure with the music, other associative and metaphorical components of the music can connect to the internal semantics. (Cohen 2013, 188)

The *associative* level is a product of the meaning we attribute to the visual content in relation to the music. As Kalinak (1992, 87) elegantly articulates it[4]:

The classical narrative model developed certain conventions to assist expressive acting in portraying the presence of emotion... close-up, lighting and focus, symmetrical mise-en-scene, and heightened vocal intonation. The focal point of this process became the music, which externalized these codes through the collective resonance of musical associations. Music is, arguably, the most efficient of theses codes, providing an audible definition of the emotion, which the visual apparatus offers... Music's dual function of both articulator of screen expression and initiator of spectator response binds the spectator to the screen by resonating affect between them.

This emotional ability of music is carried by the emotional information in harmony, rhythm, melody, timbre and tonality, which some scholars argue are a product of human evolution (Lenti Boero and Bottoni 2008). Table 1 below illustrates the binaries between internal-external semantics and congruence-associations.

---

4   Cited in Cohen (1999, 266).

Table 1: Representation of the congruence-association model

	Internal Semantics of Film	External Semantics of Film
Music Structure (Congruence)	Structural congruencies between film and music may: - engage attention (f.ex. repetition of low pitches) - direct attention to an object/subject on the screen	Music is diegetic: - Music informs the narrative, the characters,... - Music may carry humor
Music Meaning (Associations)	Music evokes emotions rooted in biological processes, e.g.: - discordant notes trigger uneasiness - Major tonality is associated with joy/happiness, minor tones with sadness	Music provides associations that help understand the narrative, e.g.: - pairing of music and film creates leitmotivs - music styles carry cultural conventions (anthems, funerals, wedding,...)

Source: Inspired by Cohen 2013,185

This model of congruence-association is internalized by the spectator and establishes a working narrative that eventually creates the prosthetic effect emphasized by Landsberg (2004). Cohen's model will constitute a central theoretical framework in our analysis of the role of music in *My Heart of Darkness*.

ELEMENTS OF MUSICAL LANGUAGE

Before we delve into an analysis of the music in *My Heart of Darkness*, we feel it necessary to review the perceptions attached to the structural musical elements we will use. These interpretations can be divided into two main categories; first, the traditional conventions giving music symbolic and literal meaning; second, cognitive studies of how people react emotionally to the different structural features in music. In many cases, these two categories overlap. We have defined four central elements: tones, dynamics, rhythm and timbre.

Tones

Cognitive studies have shown that a high pitch does not have the same effect or does not carry the same symbolic value as a low pitch. The same can be said about upwards vs. downwards movements, the former generally perceived as positive and the latter as negative (Cohen 2013, 187, Lenti Boero and Bottoni 2008). Tone intervals also play a certain role. Natural intervals like octaves and fifths are perceived as more consonant than other intervals. Likewise, large intervals are experienced as more consonant than small intervals. In the Western tradition, tonality is divided into two different modes: major and minor. Major modes have been associated with joy and happiness and minor modes with sadness and sorrow. Cognitive studies corroborate these conventions (Hunter and Schellenberg 2010, 143). A minor mode is experienced as sad, but also as more complex than a major mode: even if it sounds sad, we still appreciate listening to it. Harmony and chords are also important parts of the musical language. Major and minor triads are perceived as more concordant than augmented or diminished chords. Chords that do not require resolution sound more consonant than chords that do, for example a Major 7 chord. Furthermore, Kalinak (1992, 6) asserts that arpeggiated chords feel more unstable than chords played horizontally, i.e. simultaneously – but, as far as we know, there is no cognitive study to sustain this.

Dynamics

Dynamics also play a substantial role in music. In some studies, loudness is associated with anger, but may turn out positively if accompanied by a major key – the result can then be triumph rather than anger. Moreover, dynamics is critical in regards to emotions: tremolo is clearly associated with suspense (Kalinak 1992, 14), while a crescendo and decrescendo "help to modulate the stimulation and can be used to heighten or diminish it" (Kalinak 1992, 11). The same responses can respectively be linked to an accelerando and a ritardando, which leads us to rhythm.

Rhythm and Tempo

Along with mode, tempo and rhythm play the most prominent part in how music is perceived (Levitin 2011, 165-168), at least in cognitive studies (Hunter and Schellenberg 2010, 138). Usually, high tempo is related to happiness and low tempo with sadness. Quick tempi tend to escalate the stimulation of the nervous system, while slower tempi have the opposite effect (Kalinak 1992, 11). Syncopations and punctuations are perceived as more discordant than plain rhythms, and thus more disquieting. Dance rhythms, as well, can be very suggestive: a waltz, for example, can evoke grace and sophistication (Kalinak 1992, 13).

Timbre

Hunter and Schellenberg (2010) have indicated in their review article that timbre does not offer the same unequivocal results as the other musical elements in cognitive studies. In general, soft timbre is attached to tenderness and sadness, and sharp timbres are related to anger. Nevertheless, studies show that the combination of flute and strings in Western music is generally associated with sadness. On an external semantics level, Kalinak (1992, 13) stresses that the strings family is, in range and tone, the closest to the human voice, and thus often used to express emotions.

As pointed out by Boero and Bottoni (2008), it is probable that part of our aesthetic and emotional response to sound has been shaped by our evolutionary past. Periodicity is preferred to aperiodicity: aperiodic sounds, like those from thunderstorms or from mammalian predators, are associated with threats. The same consideration concerns pitch, timbre, and loudness: low-pitched and loud sounds are associated with those emitted by large animals or possible predators and might therefore elicit fear or anxiety in people. Large variety is more appreciated than single sounds, presumably because it evokes an abundance of species (and prey). The other part of our response has been shaped by our own experiences or is a product of the cultural context.

GENERAL CHARACTERISTICS OF THE MUSIC IN *MY HEART OF DARKNESS*

Unlike the story, the music in *My Heart of Darkness* does not stand as a whole narrative in itself, but consists of short pieces composed for different scenes. The music is composed of thirty-nine occurrences that are almost exclusively extradiegetic (37 sequences), and represents 55% of the film's length. This proportion can be considered as rather important,[5] and even more if we take into account that *My Heart of Darkness* is a documentary. Documentary is a genre where journalistic rationalism and observational minimalism have traditionally raised suspicion that music might somehow subvert the film's integrity (Corner 2005, 243). In television documentaries, for example, music has usually been employed more frequently with "lighter" topics, but it is also characteristically used in programs with a strong self-consciousness of their role as artefacts, as authored 'works' (Corner 2005, 248). This is a category *My Heart of Darkness* fits into as an autodiegetic documentary.

As a consequence, we can see a strong focalization on the autodiegetic narrator from the very beginning and throughout the whole film, something that structures the main part of the narrative. This focalization is not only a result of the focus, camera angle, and montage used in the film, it is highlighted by the prevalence of the narrator's voice-over to disclose the protagonist's inner thoughts. Moreover, the voice-over runs through the entire film, guiding the spectator through the narrator's/director's personal quest. Finally, music is used systematically in relation to the voice-over, eventually enhancing the spectator's sense of reality, shaping inferences within the discourse or carrying further meaning through its emotional load. More generally though, this focalization invites us to identify with the narrator's intimate feelings and thoughts, and music helps activate our emotional involvement and our compassion. However, by structuring the whole film, it overtly functions as a kind of *parti pris*. It creates a manifest imbalance between the omnipresence of Marius' perspective and the voices and viewpoints of the other protagonists (Samy, Patrick, and Mario). This

---

5   In a former study we have conducted on French banlieue films, music occupied an average of 60-65% of musicals, while the average for fictions was approximately 30% or 27 sequences (Wagner 2011, 320-329).

imbalance can further be associated with an unconscious colonial stance, as the contrast between the white South African main character/narrator/director and his black companions (two Angolans and a Bushman) is obvious. Whenever the film achieves a polyphonic level, by presenting the testimonies of every character regarding their war experiences, the use of music is once more very representative: there is no music at all accompanying these testimonies. How should this focalization on Marius be interpreted? It is clearly not a clear-cut Eurocentric or colonial angle. The narrator/director/main character has the best intentions and has a humble and honest aim. It is just that the film is chiefly about his own quest and story – the other protagonists do not feel the same need for redemption and forgiveness that he nurtures.

The film's music strengthens and narrows this Western perspective since it "efficiently establishes historical and geographic setting, and atmosphere, through the high degree of cultural coding" (Corner 2005, 247-248). The music in *My Heart of Darkness* clearly establishes itself within the setting of early 21st century Europe and North America.[6] The instrumentation is dominated by strings and percussions used in the western symphonic tradition. The style is, to a certain degree, minimalistic, but to this somewhat repetitive compositional technique is added a more grandiose style reminiscent of American film music, which was initially heavily influenced by the late European romantic symphonic style (Koldau 2008, 249). The two sequences of diegetic music in the movie are the only exceptions. The first one is the ringtone of Patrick's cell phone, which is a type of popular music that can be heard on cell phones anywhere in the world. The second one is the tribal music at the end of the movie that accompanies the cleansing ceremony. However, as we shall see, this tribal diegetic music is still subordinated to the extradiegetic soundtrack. This may indicate that, considering the music style, the implied spectator is chiefly a Westerner.

In addition to contributing to abolishing the frontier between the film's reality and the spectator's reality, the music of Swedish composer Jan Anderson bridges the temporal and geographical gap between a less-known

---

6 The music has many postmodern features, but also more binary opposition, tension, and finality than music from the eighties and the nineties. See Skarpeid (2015, 242-246).

war in Angola and a contemporary spectator. According to Gorbman (2003, 40), "music more or less short-circuits consciousness, it facilitates the process by which the spectator slips into the film". To produce or strengthen the emotional reception of the spectators, Anderson is employing standard composing techniques. In the following we will analyze some of the salient features of the compositions in *My Heart of Darkness*.

Notwithstanding the fact that the use of music is extensive in this documentary, the same main themes are to be found in different variations. We have isolated the six most salient moments and themes: "Main Theme" accompanies the film's opening; "Restricted Range of Effect" is the musical theme attached to the archival footage and the drawings of dramatic memories; "Childhood" accompanies Marius' testimony about his experiences in the war; "Cleansing Ceremony", towards the end of the film, functions as a kind of climax; the moment after the ritual, "Essence and Accidence", is also interesting; and finally, "The Milgram Effect" accompanies the end credits' that close the film.

SALIENT MUSICAL MOMENTS AND THEMES IN *MY HEART OF DARKNESS*

Opening

The film opens on a black screen, with a quotation from Philip Zimbardo as an epigraph, that introduces one of the film's central topics: "The line between good and evil is permeable and almost anyone can be induced to cross it when pressured by situational forces." It is accompanied by sounds of war (rounds from automatic weapons and yelling), dissonant high notes, the clear timbre of a glockenspiel playing a short figure of three notes – E-D-F in the upper register – and string instruments playing long notes in the lower register. On an internal semantic level, the lower pitch notes from string instruments played at a tempo similar to the human heartbeat engage attention and, together with some isolated dissonant higher pitch notes, evoke deep emotions associated with a sentiment of mystery and menace. On an external semantic level, the glockenspiel evokes a music box melody we might associate with toddlers, innocence, or simply time passing, while the sounds of war reinforce the feeling of menace and fear. The higher pitches of the isolated dissonant notes and of the glockenspiel create a strong contrast with the sounds of war and the lower pitch notes from a contrabass that together create an atmosphere of full-scale drama.

Moreover, the quotation sets the core of the film in an authoritative manner, as it comes from an expert, Dr. Philip Zimbardo, who has demonstrated in his research, for instance in the Stanford Prison Experiment (1971), the corrosive and negative effects high-pressure situations may have on human behavior, eventually turning ordinary people into torturers. The character font used also makes this quotation congruent with the war-like soundtrack. The white letters look like the characters the military usually employ on their shipments and materials. The first shots following the quotation are a succession of high-angle/low-angle close-ups: a high-angle close-up of a shoebox containing family photos, war photos, a dagger, a military identification tag, an ID card, and some Angolan banknotes; and a low-angle close-up of the main character's face – Marius – as he scrutinizes the contents of the box spread on the floor of his hotel room. This succession of shots produces a focalizing effect: the spectators are invited into Marius' deepest thoughts and feelings. Then comes the narrator's voice-over, with low notes from string instruments in the background. As string instruments are generally considered the closest musical translation of the human voice and human emotion, the whole combination strengthens the focalization effect and the monologue clearly heralds the central topic of the film:

What does a man have to do to regain his self worth after losing it? I've lost my family to exile, my sanity to trauma, my innocence to war. The only two things I have left in life worth living for are my two beautiful daughters. And I will surely lose them too, if I don't come to terms with my past, with my memories from the war, stuffed in a box I've been dragging with me for the past twenty-eight years.

Towards the end of this monologue a tremolo from violins intervenes (D in a two-line octave), inviting the spectator to anticipate that something is about to happen. As soon as the monologue is over, the lower strings repeat the same figure of three notes played by the glockenspiel in the beginning (E-D-F). This figure accompanies the opening credits and will function as the main theme and as a forceful leitmotiv throughout the entire documenttary. The opening credits appear, written in the same military font as the epitaph, white letters on a background of grey cloudy skies reflecting on a river in a jungle, the camera slowly moving forward. Then the waters darken and the image becomes slightly blurred. There is a rhythmic beating of drums sustained by a pizzicato from strings, and this is followed by long high-pitched violin notes, contrasting with the remaining low notes of the contrabass. We hear then the high voices of a choir coming in together with the sibylline tones of the glockenspiel, before the title of the film itself spreads out in demonstrative blood letters all over the screen. Finally, a fade out does the transition to the next scene.

The use of music here is clearly structuring – it signals the beginning of the film. Retrospectively, it is also easy to see in this opening an introduction of almost all the musical ingredients we will encounter throughout the film. Likewise, we can interpret these musical ingredients as a metaphorical transcription of some of the film's motives: the strings will invoke the complex inner voice of the main character's tormented soul, while the glockenspiel will symbolize a lost and much-sought innocence, and the choir voices a religious/divine/superior way toward forgiveness and redemption. Notwithstanding these interpretations, the opening music produces strong emotional associations. In this regard, music fulfills one of its most prominent feature in movies: it helps transport us to the world of the film, abolishing in a way the barrier between the film's and the spectator's reality. The musical analysis of the main theme will further emphasize this emotional effect on the spectator.

My Heart of Darkness "Main Theme"

As mentioned, the main theme consists of only three notes: E, D and F in D-minor, in the lower tetrachord. Even though it was first played by the glockenspiel, it is only when this leitmotiv is taken up by the strings that we clearly experience a pulse at ca. 60. It is difficult to find a particular congruence between the structural features of this theme and the structural features of the images it accompanies, but we can identify clear metaphoric associations between the images and the music, both on an internal and an external semantic level. According to Hunter and Schellenberg (2010, 138), mode and tempo are "the strongest determinants of perceived emotion in music". Here, a tempo pulsing at a normal heart rate engages deep emotions. In fact, most of the compositions in *My Heart of Darkness* have a tempo based on a pulse of approximately 60 beats per minute.

My Heart of Darkness Main Theme

Jan Anderson

In the first four bars, the same movement is repeated twice: E-D-F, played by strings. The rhythm is syncopated and punctuated. Such an irregular rhythm combined with a minor mode elicits a feeling of sadness. The repeated succession of E-D-F combined with this rhythm evokes a forceful circular movement, like a wave, a tide or a stream. The final four bars are a descending octave from G to G in four notes, with longer and consonant intervals. This, together with the combination of regular rhythm and a major mode (broken C), ending on the fourth, releases some of the tension from the first four bars and leaves the continuation open. The descending melodic figure and its rhythm resonate as a progressive fall or a profound dive. Still, the low, loud pitch sustains the feeling of uncertainty.

In addition to these internal associations, there are some external associations we would like to highlight. The timbre of the strings adds a dramatic and personal/human dimension. A single violin "may sound nostalgic and sentimental" (Neumeyer and Buhler 2001, 38). Kalinak uses the same term, timbre, and writes that "strings are often used to express emotions. In particular, the violin" (1992, 13). Kalinak sustains her argument by quoting Franz Waxman about the use of strings in the opening of *God Is My Co-pilot* (Robert Florey, 1945). He used strings because of their "deeply religious, emotional tone". The intervention of drums sustains the rhythm and the gravity of the mood distilled by the contrabass, while the pizzicato of the violins accentuates the sense of movement and the human dimension of the atmosphere. Finally, the high-pitched choir voices give the opening a heightened, almost religious character, and the glockenspiel adds a touch of innocence evoking nostalgia or childhood, as it sounds like a music box.

The repetition of the theme, promoted as leitmotiv, crystallizes the overall atmosphere of the film (Lissa 1965, 223). As expressed by Arnold Whittali (New Groove online), the leitmotiv is used "to represent or symbolize a person, object, place, idea, state of mind, supernatural force or any other ingredient in a dramatic work". Together with the narrator's thoughts presented in the voice-over, it evokes the dramatic and tormented dimension of the main character's burden and of his quest for forgiveness. Regarding the discourse on memory, the music clearly colours Marius' relationship to memory as heavy and almost haunting, lurking under the surface, in line with the image of dark river waters toward the end of the opening. The descent of the last four bars then seems to evoke a deep dive into the past, as if to unravel psychological repression in the unconscious, or a repressed memory. Emblematically, Marius' next monologue underscores this interpretation:

This box of shameful memories that all veterans carry, hidden away in their darkest closet, I have to get rid of it, once and for all, before my daughters discover it. I have to go back and reclaim my life. I have to go back and reclaim my lost innocence.

"RESTRICTED RANGE OF EFFECT"

Almost ten minutes into the film a sequence of significantly discordant music is heard. The scene is a flashback of archival footage in sepia showing the discovery of a mass grave from the Cassinga massacre. At the same time, the voice-over disserts about the banality of evil, "changing us into monsters, with a burden almost too heavy too carry".

Restricted Range of Effect

Jan Anderson

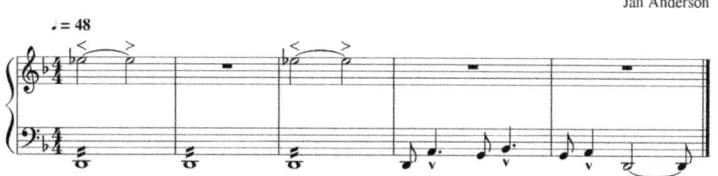

The sequence of music starts with a long low D from a contrabass. Then a long Eb from a flute is added, two octaves and a semitone higher. In addition, we can hear low tremolos from a synthesizer and the buzzing of flies. The combination of flute and strings, the discordant half tones, and the heady rhythm of the tremolos convey a strong feeling of sadness and unpleasantness/discomfort on the level of internal semantics. According to Koldau (2008, 262), the perception of the half tone interval refers to "die jahrhundertealte Assoziation der absteigenden kleinen Sekunde mit negativen Affekten lässt eine hermeneutische Deutung dieses Intervalls zu". Thereupon, a cello/contrabass plays a variation of the main theme: the first two bars are similar and then the following descent increases in tempo but remains just as low. On an external semantics level, the music is in line with the discourse presented by Marius and the horror of the pictures displayed: the minor mode, discordant half tones, tremolos, and low pitch convey a strong feeling of sadness, uneasiness, anxiety, and fear. The diversity and contrast of competing instruments playing in high and low registers we find in the opening scene is here completely absent. The expression is simpler and emotionally more plain. Interestingly, the same theme accompanies other scenes that use archives: the scene where Marius shows his war photos to the other protagonists, the pictures that constitute the materialization of his traumatic memory (54:31); and two of the three other scenes using archival footage in sepia, either to show and comment on images of UNITA leader Jonas Savimbi (at 1:08:05 in the film), or to dwell upon the serious aftermath of the war and its effect on Angolan society towards the end of the film (1:22:34). Through these emotions, the music emphasizes and helps embody the discourse and images connected to the memorial traumas of the war, presenting them as profoundly unsettling, sad, and tragic matters.

"CHILDHOOD"

Twenty-six minutes into the movie, Marius starts talking about his childhood. The music accompanying this scene matches the twelfth piece in the soundtrack, and also bears the same name. We are first shown two successive drawings, respectively of a stork bringing a child and of a child in a cradle. Then a travelling shot directs our attention to a certificate of baptism, accompanied by the same kind of classical drawings of toddlers, before it changes to traditional family pictures of Marius and friends at a very young age. The voiceover starts very early with Marius speaking of his childhood in the heartland of conservative South Africa together with white and black children. The discourse and the images differ a bit, because only white kids are seen in the pictures, but the musical expression is, at this point, very congruent with the images and the discourse. A glockenspiel plays the leitmotiv at a high pitch and is accompanied by violins in pizzicato and a triangle. The glockenspiel was played in the opening scene, and was associated with Marius' lost innocence and the picture of his two daughters. This time, the glockenspiel is clearly connected to Marius' own childhood: it repeats the main theme several times at the same pulse of 60, but at a faster rhythm, a 6/8 metre traditionally associated with the waltz (as shown in the figure below). It evokes a joyful and more spirited mood.

Childhood

Jan Anderson

However, the discourse takes another turn, first slightly incongruent with the music, and then introducing a growing ambiguity: "At the age of six, I was sent to a White African school in town. The city boys called us "kaffir boeties" – nigger lovers. And so the brain fucking started."

We then hear the ascending tones of choir voices at a high pitch – used in the opening as well – clearly referring to a sacral mode and religion, as the screen displays the cliché image of a young child praying, and of a prayer, allegedly written by Marius himself as a kid. Then we see family pictures of paratroopers and of Marius in different military uniforms, while the voice-over carries on with its explanation, overtly relating the heart of the problems to the strong religious beliefs of the Afrikaans minority:

My ancestors were Calvinists, kicked out of Holland because of their rigid attachment to the Old Testament. To the Promised Land: South Africa. We were the chosen people. Anything different was looked at with suspicion and distrust. We were taught that the Blacks were Communists, and wanted to take our land, to kill our parents and rape our women. So, getting into the army was a natural thing to do. It was expected of you. It was our destiny. It was something to defend, to fight for. But was it worth killing for?

The tirade is accompanied by the entry of strings playing a long low note that precedes the main theme (performed by a clarinet), and is followed by a crescendo of drums and low tremolo of strings. The glockenspiel and the joyfulness have faded away, yielding to a more tense and serious mood. "Childhood" lasts only two minutes and evolves from a joyful and light melody associated with toddlers' innocence to a growing ambiguity and turns finally toward a solemn and fatidic end. Regarding the memorial discourse, the music first sustains the images and the comment of the voiceover before it creates a contrast, an ambiguity, and finally a tense mood that reflects the odd and conflicting relationship Marius nurtures towards his childhood and his youth.

"Cleansing Ceremony"

The cleansing ceremony marks an important climax in the film. The protagonists go through a ritual in order to be purified and to get rid of the symbols of the war traumas they have suffered. As Marius puts it:

As we approach our final destination, Samy wants us to return to his village, to his people. They will prepare a cleansing ceremony to burn our bad memories and rid ourselves of all the evil spirits caused by the war. He wants to show that we are brothers and that it is possible to reconcile.

A goat is sacrificed in Samuel's village and the blood of the animal is used to purify the four war veterans. They burn photos and other symbolic artifacts of the war, and a party continues through the evening to celebrate these moments.

The use of the music here is interesting in many aspects. First, the soundtrack alternates between diegetic music and extradiegetic music. The ceremony opens with the diegetic: women chanting and the quick rhythm of drums, representing the music of Samuel's tribe. When the goat is about to be sacrificed, extradiegetic music takes over. A high-pitched note from violins engages our attention and sustains our expectation; a synthesizer playing an ascending and descending crescendo follows when the goat's throat is slit, stressing a feeling of lament (Rosand 1979) and, foremost of all, the importance and solemnity of the act. Violins play the first two bars of the main theme/leitmotiv in the upper register while the men are smeared with the goat's blood. The diegetic tribal music comes in again after the purification, replaced again thirty seconds later by extradiegetic music when Marius and his companions burn symbols of their war. The main theme competes with other themes played by violins and glockenspiel and sung by a choir. The sacrifice ends on the loud ascending notes of the synthesizer. Diegetic music closes the ceremony as we see a celebratory party with dancing and music playing into the night.

The alternation of diegetic and extradiegetic music is interesting, because it underlines the prominent place the extradiegetic music and the main theme take. Even when the diegetic tribal African music is used, it is subordinated to the extradiegetic Western music. However, the loudness, the high pitch, and the imbrication of different themes emphasize the gravity and importance of the scene as a capstone. The crescendo of ascending notes is congruent with the ascending flames of the fire consuming the war souvenirs on an internal semantic level. On an external semantic level, combined with the high pitch and the light tintinnabulations of the glockenspiel, the music can be associated with promise or the hope of redemption and forgiveness, a recovery of the lost innocence, evoked at the beginning, a clean slate. The joyfulness and the high tempo of the vibrant closing tribal drums and chants clearly add to this mood of relief.

"ESSENCE AND ACCIDENT"

According to Paul Ricoeur, "To follow the story is not so much to enclose its surprises or discoveries within our recognition of the meaning attached to the story, as to apprehend the episodes which are in themselves well known as leading to this end" (1984, 67). The story in *My Heart of Darkness*, starts with a wish to heal and ends with a cleansing ritual as a solution. An epilogue follows this ritual. Subsequent to the dialog is a diapraxis: the veterans help Samuel rebuild his home after a storm. This reconstruction represents a new start and symbolizes, through this concrete brotherly cooperation, the reconciliation completed in the previous scene. The musical theme used here is called "Essence and Accident" on the official soundtrack, and is played as follows:

Essence and Accident

Jan Anderson

This composition was used in the beginning of the movie, when Samuel had to leave his village. The first time, we got only portions of the composition, partly covered by the dialogues. This time, the whole piece is played, and we clearly hear the major key, a mode that enhances positive connotations. The previous musical pieces in the film are in D-minor, while this composition is in F-major. In this regard, this represents a turning point, even if, as we see in the notation above, the melodic line of the theme of "Essence and Accident" is affiliated to the "Main Theme" as it consists of the same three notes (D-E-F) and begins on the same E.

The motif E-F-D is repeated a couple of times, ending in a descending figure. It is then taken up by the strings, before a cello/contrabass steps in later on, subsequent to a choir. The composition becomes more percussive towards the end. "Essence and Accident" has almost the same pulse as "Main Theme", just a little bit faster, but still approximately at a heart rate. However, the rhythm is slightly different and, contrary to "Main Theme", there is no syncopation in "Essence and Accident". It leads to a rhythm that feels more consonant and thus is experienced as less tense. Combined with the major mode, this simpler rhythm strengthens a sense of relief and happiness. As mentioned earlier, the children's choir conveys feelings of innocence and a sacral/divine/heightened atmosphere. This echoes the final sentences of Marius' voice-over, in which he conveys a feeling of relief and peace as he seems reconciled with his past after the journey he has performed with Patrick, Samuel, and Mario.

The dynamics are relatively strong and somewhat resume where the ritual ended, but the intensity weakens gently and a bass ostinato is introduced at the end of the piece: F#, A, D. Conforming to Cohen (2013, 185), for whom "a low repeating pulse ... engages subcortical processing and higher brain levels", this repetitive figure at the end can be apprehended as the forewarning of the disturbing harmonic ending of the piece. The composition concludes on the dominant, a chord giving rise to tension and requiring release. The expected tonic remains absent and the piece thus ends on this uncomfortable tension. To a certain extent, the explanation comes immediately after: in the following scene, Patrick and Samuel role-play (like children) their respective part in the battle of Cuito Cuanavale, simulating the shouting and sounds of war.

Unlike the other pieces, instead of a quiet and low start, "Essence and Accident" begins loudly and gradually declines slightly. The melody is minimalistic and consists more of harmonies than of a proper melody. Anderson uses mostly the same instruments and elements as before. Notably, the choir instills the same sort of sacral atmosphere. However, the most striking feature is, together with the volume of the introduction, the descending bass line in major that marks a clear counterpoint to the healing ritual. As such, the music here clearly supports the idea of transition, a new start in the protagonists' relationships to their war trauma and negative memories.

"THE MILGRAM EFFECT"

Emblematically, the piece of music accompanying the end credits is called "The Milgram Effect", referring to the eponymous experiment performed in the sixties: this is here clearly placing Marius' posttraumatic stress disorder and the protagonists' problems of guilt within the frame of the Milgram Experiment. This works as a way of definitively clearing the burden of guilt and responsibility, implying that the veterans have been the victims of a broader system and circumstances that forced them to obey orders and to perform acts in conflict with their personal conscience. As this music accompanies the end credits, it works as a kind of conclusion to the film.

The Milgram Effect

Jan Anderson

The music is as usual minimalistic and the theme is composed of a short figure that is repeated, the first four bars, before it is released by a new figure on the next four bars. The eight bars are then reiterated many times. The atmosphere is neither sad as in the "Main theme", nor joyful as in "Essence and Accident", but rather somewhere in-between. This in-betweenness is labeled concurrently by the rhythm, the pulse, and the tempo, as well as by the timbre and the tonality of the piece.

Even though the first figure is different from the "Main Theme", there is an allusion to it, at least rhythmically. Like the main leitmotiv, the first figure here consists of two long tones separated by two short ones, and is syncopated. The leitmotiv of the "Main Theme" is both syncopated and punctuated, which in comparison renders the first figure here a little more concordant. Furthermore, the rhythm in bars 5-7 sounds like a sort of dance, and this movement is underscored by the pizzicato in the accompaniment. Nonetheless, the pulse is slower (48), which has a lulling effect, while the tempo of the contrabass and the violin in the accompaniment is swift and implies hastiness.

Moreover, the timbre also has a decisive impact. A flute plays the first four bars, while a violin performs the next four. As we already mentioned, the combination of flute and violin is often associated with sadness (Hunter and Schellenberg 2010, 138). However, this sorrowful mood is partly balanced by the triangle and the glockenspiel.

Finally, "The Milgram Effect" mixes minor and major modes: the first four bars are in D Minor, while the next four ones are played in F Major. This transition conveys a lighter and less melancholic atmosphere before it goes back to the first four bars. However, there are few harmonic cadences: the tonic D is missing in the first four bars; instead, the fifth is the emphasized, and the short tones at the end of meter 1 and 3 lead again to the dominant. The first figure stands as a possible ending sequence, but the composer crushes this expectation of an ending by adding other sequences, while the lower line of the contrabass, in D, promises that the upper line will at last return to the tonic.

The Milgram Effect

Jan Anderson

This illustrates how the theme of "The Milgram Effect" blends concordance and discordance, thus being not as discordant (sad and dramatic) as the "Main Theme" leitmotiv, but not as concordant (light and happy) as "Essence and Accident" either. "The Milgram Effect" conveys ambivalent feelings, a bittersweet mood, neither completely somber nor joyful. In relation to the film's discourse about memory, it exhibits, in this musical way, two different sides. On the one hand, a bright and happier tone, light and almost childish, unworrying and smiling. On the other hand, we do not find the heavy, drastic and fateful figure of the "Main Theme", though a bit of it remains. This can be seen as a new state of mind for the four protagonists – especially the main character Marius – after the end of their quest. They are partly reconciled with themselves and the bad memories but not completely rid of them. This seems reflected by the end of the piece on the soundtrack.

Towards the end of "The Milgram Effect", a progressive liquidation takes place, where most of the musical elements are removed. First the bass disappears, then the glockenspiel, then the pizzicato strings, and in the last four bars, the triangle. We are left alone with a harp and a violin (see the notation above). The harp plays eighth notes as accompaniment. The violin plays both the upper line and the melody line using flageolet technique. The melody does not end on the tonic, but on the fifth, thus producing an open chord. In addition, the third is absent: the chord is neither major nor minor, which strengthens the feeling of an unbiased ending, neither glad nor sad. In other words, the last chord is left open, literally pending, leaving the spectator free to interpret the resolution of the film. This openness could also signify that reconciliation with traumatic memory is a complex process that is never fully achieved.

Conclusion

The important space occupied by music in *My Heart of Darkness* contributes significantly to this documentary's presentation of itself as a self-conscious authored work in line with its autobiographic character. If we attempt to read the rhetoric of memory in this film according to Erll's model (2008), at least three of the four modes are present: the experiential, the mythical and the reflexive modes. One could even assert that the antagonistic mode is also an undercurrent in the overt victimization of the protagonists when it comes to their partaking in the war. Music functions significantly in this documentary as it increases the emotional impact of the scenes it accompanies, most notably Marius' voiceover discourse on memory. Music highlights the dramatic character of the war trauma and the crucial existential aspect of the reconciliation process at work. Combined with a focalization through the main character of Marius, music transcribes and embodies what is at stake in Marius' and his companions' search through this trip. However, the prominence of the music also contributes to a fictionalization of this documentary and to an overfocalization through Marius, at the expense of the other protagonists. Paradoxically though, the scenes (without music) where they express themselves loom even more markedly.

By using Cohen's congruence-association model and interpretations related to traditional conventions and to the reception of musical elements like tones, dynamics, rhythm and timbre, we hope to have shown in detail how the salient musical moments can be analyzed and interpreted in the frame of the memorial discourse of the film. The "Main Theme", with its minimalistic three notes in a minor mode and its irregular rhythm, elicits sadness and profoundness. As it is closely associated with Marius, it evokes the dramatic and tormented dimension of his quest to handle his haunting memories. "Restricted Range of Effect" is used in relation to archival footage and photos, and underscores, through its discordant elements, the horrors of the war. "Childhood" expresses first naïve innocence and happiness but evolves towards sorrow, symbolizing Marius' youth as a fraud or a lost paradise. "Cleansing Ceremony" marks a climax where the protagonists are allegedly purified of their sins and war traumas, as the high pitch and tempo and the joyfulness of the music evokes a feeling of relief and a hope for redemption. After the ritual ceremony, "Essence and Accident" prolongs this sentiment and supports the idea of a new beginning for the film's main characters. But the theme accompanying the end credits, "The Milgram Effect", in its mix of major and minor modes, its concordance and discordance, introduces ambivalence into any interprettation of the final result of the characters' quest.

In general, it is interesting to notice that music evolves from themes in minor modes that are dark and disquieting to final themes in major modes that are more joyful. These final themes invokes the possibility of hope, something missing in the leitmotiv. However, the end credits' theme nurtures an ambiguity by mixing contrary musical elements and by closing on an open chord, leaving the end open for interpretation. In this regard, music conveys the idea that even if Marius has found some relief in his quest, a complete reconciliation with his past remains a complex and long process, still to be negotiated.

It is well possible that after having heard the music of this film that much, we have become mesmerized by it beyond all measure. However, we argue that music brings to the film a great emotional power that strengthens and sometimes surpasses the discourse and the images. It contributes to the film's discourse on memory by supporting and reinforcing it, but also by adding to it a certain subtlety and a dose of nuance. In this regard, Anderson's music, through its cleverness and its elegance, appears to us to be a masterpiece.

REFERENCES

Baumgartner, Thomas, Kai Lutz, Conny F Schmidt, and Lutz Jäncke. 2006. "The emotional power of music: how music enhances the feeling of affective pictures." *Brain research* 1075 (1):151-164.

Bezdek, Matthew A, and Richard J Gerrig. 2008. "Musical emotions in the context of narrative film." *Behavioral and Brain Sciences* 31 (05):578-578.

Bolivar, Valerie J, Annabel J Cohen, and John C Fentress. 1994. "Semantic and formal congruency in music and motion pictures: Effects on the interpretation of visual action." *Psychomusicology: A Journal of Research in Music Cognition* 13 (1-2):28.

Boltz, Marilyn G. 2001. "Musical soundtracks as a schematic influence on the cognitive processing of filmed events." *Music Perception* 18 (4): 427-454.

Boltz, Marilyn G. 2004. "The cognitive processing of film and musical soundtracks." *Memory & Cognition* 32 (7):1194-1205.

Boltz, Marilyn, Matthew Schulkind, and Suzanne Kantra. 1991. "Effects of background music on the remembering of filmed events." *Memory & Cognition* 19 (6):593-606.

Bondebjerg, Ib. 2014. "Documentary and Cognitive Theory: Narrative, Emotion and Memory." *Media and Communication* 2 (1):13-22.

Bordwell, David, and Kristin Thompson. 1979. "Sound in the Cinema." In *Film Art: An Introduction*. Reading, MA: Addison-Wesley.

Bullerjahn, Claudia, and Markus Güldenring. 1994. "An empirical investigation of effects of film music using qualitative content analysis." *Psychomusicology: A Journal of Research in Music Cognition* 13 (1-2):99.

Burgin, Victor. 2004. *The remembered film*. London: Reaktion books.

Cohen, Annabel J. 1999. "The functions of music in multimedia: A cognitive approach." *Music, mind, and science*: 53-69.

Cohen, Annabel J. 2001. "Music as a source of emotion in film." Music and emotion: Theory and research:249-272.

Cohen, Annabel J. 2013. "Film Music and the Unfolding Narrative." InLanguage, Music and the Brain (Strüngmann Forum Reports), ed. Michael A. Arbib.

Corner, John. 2005. "Sounds Real: Music and Documentary." In *New challenges for documentary*, edited by Alan Rosenthal and John Corner, 242-252. Manchester University Press.
Desgranges, Béatrice, and Jean Eustache. 2010. *Les chemins de la mémoire*. Paris: Editions Le Pommier.
Dickinson, Kay. 2003. *Movie music, the film reader*: Psychology Press.
Ekman, Paul. 2007. *Emotions revealed: Recognizing faces and feelings to improve communication and emotional life*. New York: Henry Holt and company.
Ekman, Paul, and Wallace V Friesen. 1975. *Unmasking the face: A guide to recognizing emotions from facial clues*. Englewood Cliffs, NJ: Prentice-Hall.
Erll, Astrid. 2008. "Literature, film, and the mediality of cultural memory." *Cultural Memory Studies: An International and Interdisciplinary Handbook* 8:389.
Erll, Astrid. 2010. "Literature, Film, and the Mediality of Cultural Memory." In *A companion to cultural memory studies*, edited by Astrid Erll and Ansgar Nünning, 389-398. Berlin: De Gruyter.
Erll, Astrid. 2011. *Memory in culture*: Palgrave Macmillan.
Gorbman, Claudia. 1987. *Unheard Melodies: Narrative Film Music*: Indiana University Press.
Gorbman, Claudia. 2003. "Why Music? The Sound Film and its Spectator." In *Movie Music The Film Reader*, edited by Kay Dickinson, 37-48. London: Routhledge.
Hoeckner, Berthold, and Howard C Nusbaum. 2013. "Music and Memory in Film and Other Multimedia: The Casablanca Effect." *The Psychology of Music in Multimedia*: 235-66.
Hunter, Patrick G, and E Glenn Schellenberg. 2010. "Music and Emotion." In *Music Perception*, edited by Mari Riess Jones, Richard R. Fay and Arthur N. Popper, 129-164. New York: Springer.
Juslin, Patrik N, and John A Sloboda. 2001: *Music and emotion: Theory and research*: Oxford University Press.
Juslin, Patrik N, and Daniel Västfjäll. 2008. "Emotional responses to music: The need to consider underlying mechanisms." *Behavioral and brain sciences* 31 (05):559-575.
Kalinak, Kathryn. 1992. *Settling the Score: Music and the Classical Hollywood Film*. Madison, WI: The University of Wisconsin Press.

Koldau, Linda Maria. 2008. "Kompositorische Topoi als Kategorie für die Analyse von Filmmusik." *Archiv für Musikwissenschaft* 65 (4):247-271.

Kuhn, Annette. 2002. *Family secrets: Acts of memory and imagination*. London & New York: Verso.

Landsberg, Alison. 2004. *Prosthetic memory: The Transformation of American Remembrance in the Age of Mass Culture*. New York: Columbia University Press.

Larsen, Peter. 2013. *Filmmusikk. Historie, analyse, teori*. Second ed. Oslo: Universitetsforlaget.

Lenti Boero, Daniela, and Luciana Bottoni. 2008. "Why we experience musical emotions: Intrinsic musicality in an evolutionary perspective." *Behavioral and Brain Sciences* 31 (05):585-586.

Levitin, Daniel J. 2011. *This is your brain on music: Understanding a human obsession*: Atlantic Books Ltd.

Lissa, Zofia. 1965. *Ästhetik der Filmmusik*. Berlin: Henschel.

Markowitsch, Hans J. 2010. "Cultural Memory and the Neurosciences." In *A companion to cultural memory studies*, edited by A Errl and Ansgar Nünning, 275-283. Berlin: De Gruyter.

Marshall, Sandra K, and Annabel J Cohen. 1988. "Effects of musical soundtracks on attitudes toward animated geometric figures." *Music Perception*: 95-112.

Meyer, Leonard B. 1956. *Emotion and meaning in music*. Chicago: University of chicago Press.

Montepare, Joann M, Sabra B Goldstein, and Annmarie Clausen. 1987. "The identification of emotions from gait information." *Journal of Nonverbal Behavior* 11 (1):33-42.

Neumeyer, David, and James Buhler. 2001. "Analytical and interpretive approaches to film music (II): Analysing Interactions of Music and Film." In *Film music: Critical approaches*, edited by K. Donnelly, 16-38. Edinburgh: Edinburgh University Press; Continuum.

Radstone, Susannah. 2010. "Cinema and Memory." In *Memory: histories, theories, debates*, edited by Susannah Radstone and Bill Schwarz, 325-342. Fordham Univ Press.

Ricoeur, Paul. 1984. *Time and Narrative Volume 1*. Translated by Kathleen McLaughlin and David Pellauer. 3 vols. Vol. 1. Chicago: Chicago University Press.

Rosand, Ellen. 1979. "The descending tetrachord: an emblem of lament." *The Musical Quarterly* 65 (3):346-359.

Skarpeid, Jon. 2015. "Kosmologi i miniatyr? Narrativitet i hindustanimusikk sett i relasjon til indisk religion." PhD, NTNU.

Tan, Siu-Lan, Matthew P Spackman, and Matthew A Bezdek. 2007. "Viewers' interpretations of film characters' emotions: Effects of presenting film music before or after a character is shown." *Music Perception: An Interdisciplinary Journal* 25 (2): 135-152.

Vitouch, Oliver. 2001. "When your ear sets the stage: Musical context effects in film perception." *Psychology of Music* 29 (1):70-83.

Wagner, D.A. 2011. *De la banlieue stigmatisée à la cité démystifiée: la représentation de la banlieue des grands ensembles dans le cinéma français de 1981 à 2005*: Peter Lang.

"I don't trust in pictures"
Forms for Authentication in *My Heart of Darkness* and Annekatrin Hendel's *Vaterlandsverräter*

BENEDIKT JAGER

TEUTONIC OVERTURE

In a shorter documentary film from 2011 – *Flake* – Annekatrin Hendel portrays the keyboardist of the German rock band Rammstein. This band has its origins in the former GDR and Christian Flake Lorenz has been a part of the legendary punk-band "Feeling B". Rammstein is a very successful band abroad and still questionable in Germany because of their use of a very German iconography to allude to fascistic iconography. Their shows and videos contain elements that remind one of Wagner operas and Teutonic mythology. It is not hard to see that the world of Rammstein is highly artificial. In this artificial universe, the keyboardist Christian Flake Lorenz is used as the irony-marker. This tall, very thin guy can be seen as a challenging the aggressive and muscle-bursting bodies of the rest of the band. When the director ask him in the film why Rammstein is so successful, his answer is very concise: "Ich denke es liegt wirklich dran, dass die Leute so ne Authentizität merken und merken sie sehen was, was echt ist und was nicht ausgedacht ist... dat wir wirklich meinen was wir sagen und nicht stumpf Erfolg haben wollen, dass es uns ernst ist. Das gibt's nicht viel so auf der Welt, dass man noch watt macht weil man datt auch so meint."[1] The question of authenticity is always central in documentary films but it seems that Annekatrin Hendel is specifically triggered by this complex.

MY HEART OF DARKNESS ON THE EDGE OF TOWN – AUTHENTICITY

Here is not the place to give a sufficient history or definition of the term authenticity[2] but some remarks are necessary to establish a field of understanding. As the history of the concept shows, the excessive use of the word is quite a new phenomenon and can be dated to the second part of the twentieth century. Today authenticity is largely part of branding-strategies and anything can be labled as authentic: food, clothes, holidays and of course works of art. The high priest of authenticity in pop culture is maybe "The Boss", Bruce Springsteen. His concerts are governed by an impetus that could be called "anti-mise-en-scène", in opposition to Rammstein with their references to opera. Working hard (playing long concerts), sweating, and the use of conventional show effects that don't break the conventional frame, are ingredients in his staging that, of course, also have to be characterized as staging or a "mise-en-scène".

The fact that the word authenticity "has become part of the moral slang of our day points to the peculiar nature of our fallen condition, our anxiety over the credibility of existence and of individual existences", as Trilling (93) has noted. Most of the researchers see this increasing focus on authenticity as signalling a crisis that also might function as its own solution:

Die große Karriere des Authentizitätsbegriffs in der zweiten Hälfte des 20. Jahrhunderts scheint darin begründet zu sein, daß der Einzelne in seiner Rolle als

---

1 Flake. Meine Leben. Directed by Annekatrin Hendel. Berlin: itworks. 2011b 31.50
2 You can find a profond introduction by Knaller/Müller: Authentizität (2006). Another book that sums up the axiom is Lionel Trilling's Sincerity and Authenticity from 1971. This work does of course not reflect the post-structuralist critique of the concept of authenticity.

allgemeiner Mensch, als besonderes Mitglied von Gemeinschaften, Organisationen und als unvergleichliches, singuläres Individuum keinen Ort mehr findet, von dem aus er in der 'obdachlosen' Moderne diese unterschiedlichen Zumutungen zu synthetisieren vermöchte. So gesehen ist der Authentizitätsbegriff Ausdruck und zugleich Symptom dieser Krise, also ein Krisenbegriff, der die Krise erfaßt, selbst stets in der Krise ist und zugleich als 'Zauberwort' die Krise zumindest partiell unsichtbar macht. [...] Nicht umsonst ist der Nachbarbegriff zu Authentizität Autorität und in nicht allzu ferner Nachbarschaft wohnen Aura, Charisma und Souveränität. (Knaller/Müller, 10f.)

The transcendental homelessness that Lukács saw as a sign of modern times is visible in both films this article treats. In Van Niekerk's case – the main chrakter of *My Heart of Darkness* – the biographical situation of exile and the reference to trauma both clearly figure in the decentring of the subject. In Gratzik's case – the main character of *Vaterlandsverräter* – it is the historical failure of the socialist project, also known as the German Democratic Republic, that led to his marginalization in the cultural life of the new Germany after 1989. The experience of loss are significant for both of them and the issue of nostalgia and longing is therefore at stake: "Authenzität als moderne Kategorie verweist außerdem auf das Begehren nach Echtheit Unmittelbarkeit von Erfahrungen und Erlebnissen, nach kohärenter Identität und Ursprünglichkeit" (Daur, 2014:9f.). Here it is necessary to point out crucial differences between the two films. While Hendel's Vaterlandsverräter must be characterized as classical documenttary – a director asking a person questions and making personal decisions (paradigmatical and syntagmatical) – the situation in *My Heart of Darkness* is slightly different. Van Niekerk is not only the object of the film, he is also the co-editor together with Staffan Julén. The project therefore blurs the lines between documentary and autobiographical film-essay. Because of this co-editorship, nostalgic and sentimental traits are more apparent in *My Heart of Darkness*.

In this way the idea of "authenticity" has been rejected in several ways. For example, Niklas Luhmann described the concept as "Kultform der Naivität" and disqualified it from having any heuristic power. (Knaller/ Müller, 9.) And Andreas Huyssen argues that authenticity is only available as a ruin that is related to the special condition of Western modernity. He stresses that the discourse of authenticity is infected by doubts and therefore tends to mystifications. Authenticity is projected back in history, as in the case of the ruin, or it is connected with other constructions that are beyond discursivity. First of all, Huyssen points out the focus on subjectivity and individuality arguing that the personal experience cannot be surmounted. But this experience can most often be confronted with another subjectivity which raises the problem of intersubjectivity. To avoid these conflicts Huyssen describes the postmodern affection for trauma as a loophole: "[...] Trauma als ein nicht hintergehbar Authentisches, das nur im witness-Diskurs zu umkreisen, aber nie festzumachen ist. Authentizität nur noch zugänglich in der Katastrophe und letztendlich ohne subjektive Erfahrung [...] auch dies ein Teil der Ruinenimagination." (Huyssen, 236) These constructions shall function as an Archimedes point in the "transcendental homelessness".

Against the backdrop of these discussions, this paper is not interested in authenticity in a moral way, judging if the films are authentic in the sense of truth or reference, but wants to examine the longing after authenticity, the fictions of authenticity, the effects of authenticity, and foremost the strategies that create the illusion of authenticity (see Daur, 2014:10). The term "mise-en-scène" [Inszenierung] is therefore central in this approach. Erika Fischer-Lichte writes about the concept: "In diesem Sinne meint der Begriff Kulturtechniken und Praktiken, mit denen etwas zur Erscheinung gebracht wird. Entsprechend wird Inszenierung als ein ästhetischer bzw. ästhetisierender Vorgang begriffen und ihr Resultat als ästhetische bzw. ästhetisierte Wirklichkeit – zum Beispiel in den Bereichen Politik, Wirtschaft, Verhalten, Natur" (Fischer-Lichte, 2007:19). Some aspects in this quotation seems obvious, but one aspect is perhaps so obvious that we miss it. Something absent is brought to appearance and the project of authentication has entered the realm of reflexivity. In pretending to deliver "the real thing", these projects are caught in an ontological contradiction. The work of art (fiction or documentary) gives, of course, no access to the real thing. Reflecting or representing authenticity destroys the whole project of being authentic. The imperative "Be authentic" generates an endless regress of introspection and at last inauthenticity. In other words, "Zeigend vermag man also die ‚illusionäre', suggestiv erzeugte Identität von Signifikant und Signifikat (Unmittelbarkeit) zu präsentieren, beuge ich mich nachträglich darauf zurück, ist es unmöglich die Differenz von Signifikant und Signifikat zu verdecken" (Knaller/Müller, 9).

My Heart of Darkness

As already indicated these films treat very different subjects: Van Niekerk and Julén tell a story of the Angolan civil war while Annekatrin Hendel portrays a person of some importance in the cultural field of the GDR. Paul Gratzik was a well known writer in the GDR who wrote for very different audiences. Beside his production of fiction and texts for the stage, he wrote "nonfiction" to a very special addressee: the Stasi, the East-German secret police. It is obvious that topics such as betrayal, deception, or fraud are central. In the world of the secrets service and of spies, the question of authenticity is always a matter of significant interest. Of course, it is only a short step from that to the questions of reconciliation and forgiveness that are central in Van Niekerk's project.

There are also other striking differences between these films in terms of their modes of narration. Annekatrin Hendel's film documents a historical person, while Julén and van Niekerk's film is close to an autobiographical project. Van Niekerk has written the screenplay taking up his own biographical history that led him from South-Africa via his engagement in the Angolan civil war to his exile in Sweden. Julén and van Niekerk's connection as co-directors is not broached in the film, but van Niekerk's contribution is implicit as he fills the role of the narrator in the film's voiceover and addresses himself as "I" and "me". In opposition to this, Hendel represents her involvement in the stories in very few sequences. She explains that she grew up in the GDR and was only very peripherally in touch with these celebrities of the East German cultural sphere.[3]

Does this suggest that linking these films in this way is arbitrary? Beyond forgiveness and reconciliation the two projects are both grounded in a postcolonial situation: the white former paratrooper searching for contact with former enemies and a writer who had to define himself again after the downfall of the GDR. Both struggle with a problematic, haunting past.

---

3   Vaterlandsverräter. Directed by Annekatrin Hendel. Berlin: itworks 2011. 4:00 + 1:34:54

Men in Boats

The title *My Heart of Darkness*, with its intertextual allusion to Joseph Conrad's classical novel, indicates that Julén/van Niekerk see this journey in a boat up a river as a central part of the film's plot. After a short prelude briefly introducing van Niekerk's trauma, the camera hovers over a river in the evening light as the opening credits appear (1:52-2:58). The use of music (see Wagner/Skarpeid in this book) and the use of the warm light create a melancholic, almost melodramatic mood. When the narrator says about his two daughters, "I will surely lose them too, if I don't come to terms with my past" (1:32-1:38), the spectator understands that the boat trip is presented from van Niekerk's perspective as an existential journey. Going up the Kwando river in Angola is a travelling back in time, and the constellation river and boat can be seen as a Bakhtinian chronotope, a configuration of time and space revealing a perspective and an ideology concerning life and selfhood.[4] Everything on the way back to the battlefields reflects van Niekerk's introductory wish to regain "self-worth", "innocence", and "sanity". The river of life indicates a linearity that can be reversed. The possibility of a fundamental healing is not questioned.

Hendel also starts her film *Vaterlandsveräter* in a boat. Before the first shot, the soundtrack evokes a summer day with twittering birds and the hum of bees with underlying sounds that can later be identified as oars. Paul Gratzik, in a white shirt with a tie and a brown old-fashioned hat, rows on a small lake with Annekatrin Hendel who is invisible in the back of the boat.

---

4   "Der Chronotopos bestimmt die künstlerische Einheit des literarischen Werkes in dessen Verhältnis zur realen Wirklichkeit. Daher schließt der Chronotopos im Werk immer ein Wertmoment in sich ein, das aus der dem Ganzen des künstlerischen Chronotopos nur in abstrakter Analyse herausgelöst werden kann. In der Kunst und Literatur sind alle Zeit- und Raumbestimmungen untrennbar miteinander verbunden und stets emotional-wertmäßig gefärbt. [...] Kunst und Literatur sind durchdrungen von chronotopischen Werten unterschiedlichen Grades Und Umfanges. Jedes Motiv, jedes gesonderte Moment eines Kunstwerkes ist ein solcher Wert." (Michail Bachtin 2008:180. Chronotopos. Frankfurt a.M.: Suhrkamp Taschenbuch Wissenschaft.)

The setting of the first interview in the film is also highly loaded with connotations and this boat trip is a recurrent pattern in the film though it does not encourage a sense of linearity. While *My Heart of Darkness* starts with a monological voiceover by van Niekerk, *Vaterlandsverräter* starts with a question posed by the interviewer. Paraphrasing: How could you write reports about people you were very close to? Gratzik answers that he has only done it once because he has internalized what his mom said to him when he was a child: "'Der größte Feind im ganzen Land ist und bleibt der Denunziant.' Klar? Und dieses Wort ist nie aus mir rausgekommen. Das hat immer in mir genagt." (Hendel, 2011: 1.13). Hendel takes up that formulation and asks him to deepen it. Gratzik's reaction is quite surprising and can only be described as a U-turn. He gets obsessed and shouts that nothing gnaws inside him and that she must stop immediately, otherwise he will go overboard: "Ich hör die Scheiß westdeutschen Filmfragen genau raus, glaub mal nicht" (Hendel, 2011: 1.42). He gets more and more obsessed and works himself up into tirade about the capitalistic order. This culminates in the vision of personally becoming part of a revolutionary movement, throwing grenades at the Thyssen and Deutsche Bank buildings. "Jetzt riecht's nach Pulverdampf. Und die Stunde kommt! Und hoffentlich ohne Blutvergießen ..." (Hendel, 2011: 3.29). His speech at this point is obviously incoherent and can be read as a sign of a person struggling with great inner contradictions. Gratzik is rowing this whole time and has to correct his directions several times. We see no goal for this unpleasant trip and, in fact, Gratzik is going round in circles, unable to escape his past.

The boat and the boat journey as chronotope for life and society in the two films implies different notions of selfhood. These differences have consequences for the mediation of the journey and most of all for the status of central documents. Both films uses a wide range of different documents such as private photos, pictures from newspapers, and documentary film sequences. They are mostly used as illustrations of van Niekerk's voiceover narration or to accompany the explanations from people close to Gratzik. Let us first have a look at the notions that accompany the boat trips. The material is more extensive in *My Heart of Darkness* because the journey up the Kwando river is the backbone of the whole narrative. In the beginning of the film, it is undeniable that the journey up the river is an inner journey. In the voiceover narration, Marius declaims: "Slowly, slowly we creep up the big brown snake. This flood of water once filled with the bodies of our enemies. The blood. Your blood and mine. Together we move as former enemies. What we share lies there ahead of us, within us. There in the darkest of the forest" (Julén/Niekerk 2010:9:38). The narration is not chronological but the viewer understands that the moving forward is parallel with a moving backwards. Moving up the river is going back to the origins of the former soldiers' lives and traumas. The chronology of the movie is therefore not a problematic one and fits into van Niekerk's understanding of linearity.

The looking back into the past of the four former enemies – Samuel, Patrick, Mario, and Marius – is placed in an interesting setting. It is not a break with the realistic code of the film to collect these four people around a campfire on the shore of the river. In this setting, the film gives a location for their histories to be told. How they got involved in the civil war, how they experienced their first battles, and how war still influences their lives. First Marius asks questions, but then they each tell their own stories without interruption by the leader of the expedition. The fire in the night out in the African Bush evokes a topic that Wolfgang Kayser called an "Ursituation des Erzählens"[5] and it is not farfetched to connect this situation with Walter Benjamin's characterization of the "storyteller". In the eponymous essay from 1936, Benjamin describes the storyteller as a disappearing figure in modernity. Due to the changes in the modern world, the novelist does not have the same authority as the ancient storyteller:

"Seen in this way, the storyteller joins the ranks of sages. He has counsel – not in a few situations, as the proverb does, but for many, like the sage. For it is granted to him to reach back to a whole lifetime […]. His gift is the ability to relate his life, his distinction, to be able to tell his entire life. The storyteller: he is the man who could let the wick of his life be consumed completely by the gentle flame of his story. This is the basis of the incomparable aura about the storyteller […]. The storyteller is the figure in which the righteous man encounters himself."[6]

---

5   Wolfgang Kayser: Das sprachliche Kunstwerk. Bern: Francke Verlag 15. Aufl. 1971, 349.
6   Walter Benjamin: The storyteller. Reflections on the works of Nicolai Leskov. From Hale, Dorothy J, Ed. The Novel: An Anthology of Criticism and Theory 1900-2000. Malden, Mass.: Blackwell Publishing, 2006. 378.

This idea is very important for the film's strategies of authentication. The primal scene of storytelling is characterized by the direct contact between the group members. This setting minimizes the effects of later forms of mediation (writing, books, films, etc.) that are marked by distance and absence. The danger of misunderstanding that is inherent in the later forms of storytelling are avoided. To be more precise, the setting gives the illusion of communication without interference. The film is in these parts an illustration of what Derrida has criticized as "Phonologocentrism"[7]. Placing the four men around the fire with all the archaic notions associated with this scene, the film whispers to us that the spectators now get access to the "real". Furthermore the asymmetry of the communication is played down and underlines the existential vigour of the project.

An interesting scene that emphasizes this aspect is when Marius is retelling a horrible story about his involvement in the war. He explains the situation by drawing a sketch of the situation. This is shot in full sunshine. When the scene reaches its climax – Marius explaining the problems that occurred cutting the throat of a dying enemy with his army knife – the scene cuts over to the campfire (Julén/van Niekerk 47:00). The soundtrack of his voice does not change and one wonders it was recorded later. This transition exposes the notions mentioned above. This strategy converges with other elements in the film that pull the ideology of the film in the same direction. The slow rhythm of the film's representation of travelling up the river offers many occasions to integrate pictures that mirror the travelers' points of view. This, of course, a possibility for identification – the spectator perceives the same things as the four soldiers. What does the film show of the surroundings? It seems to be impossible to evade the maelstrom of the exotic. In recurring sequences, the animal life of Angola, hippopotamuses, elephants, and crocodiles (see Julén/van Niekerk 2010: 43:00), are portrayed as escorting the vessel and must be understood as messengers: you are leaving the (corrupted) cultural zone and entering the natural (pure, innocent) sphere.[8] Another facet of this return to the origin is

---

7   See Jacques Derrida: Grammatologie. Frankfurt a.M.: Suhrkamp Taschenbuch Wissenschaft 4. Aufl. 1992.
8   Lionel Trilling argues that imaginations of the authentic often end with the primitive: "Certain exemptions are made: the poor, the oppressed, the violent, the primitive. But whoever occupies a place in the social order in which we

illustrated by young African boys paddling in a dugout on the river. This return has to be seen as an inversion of the fall of man. In a later sequence, this topic is picked up by the narrator: "To place such destruction on the fingertips of innocent young boys. Young angels of death killing in God's name." (Julén/Niekerk 2010:44.36). The boat trip becomes a secular pilgrimage and the goal is the absolution of sins.[9]

In some way the argumentation of this paper can be perceived as questionable, but there is also a campfire scene in *Vaterlandsverräter*. The contrast could not be more extreme, from the warm Angolan jungle to the snowy winter landscape of the Uckermark in Brandenburg close to the farm where Paul Gratzik lives under quite miserable circumstances. From the community of four men to a solitary forgotten East-German writer, Gratzik sits "alone" beside the fire, answering Annekatrin Hendel's questions. This scene is a re-enacting of a previous bonfire. Gratzik is going back to one of his favorite places where himself, Heiner Müller, Steffie Spira and Fritz Marquardt[10] – members of the aristocracy in the GDR theatre – once were gathered around a campfire. The differences are striking, but there are also similarities: Even if Gratzik is sitting there now alone and is commenting on the irreversibility of time, he actualizes the notion of community. The relationship to the (now canonized) Heiner Müller is described as one of brothers, while Gratzik allocates him the role of the elder, the caring brother. These are important issues for Gratzik because of how he entered the cultural sphere in the GDR. He came to writing via the writing workers movement and as an autodidact, his position has been precarious. On the one hand, he was an example for what the authorities formulated in the

---

ourselves are situated is known to share the doom [to be inauthentic]." (Lionel Trilling (1971:102) Sincerity and Authenticity. Cambridge MA: Harvard University Press.102).

9   The film is taking up religion as a reference in describing van Niekerk's family context. The heritage from Calvinism (Julén/Niekerk 2010: 26.57).
10 Heiner Müller (1929-1995): His relationship to the Stasi is still debated today; Steffie Spira (1908-1995) very popular actress in the GDR; Fritz Marquardt (1928-2014) German director and actor connected to the Volksbühne in Berlin and to the Berliner Ensemble.

Bitterfeld program – "workers climbing the heights of culture".[11] On the other hand, he was a stranger among colleagues, which already had allocated cultural capital. Standing beside the fire, Gratzik reveals his longing for that community and mirrors his position in the reunified Germany. Like many other writers from the GDR, his reputation has waned, not to mention his financial situation.[12]

While Gratzik's fire captures this double bind, the campfire scenes in *My Heart of Darkness* seems less ambivalent. Contrary to Gratzik, who is talking to the interviewer and perhaps mostly to himself, the situation of communication is very different for Marius van Niekerk. His outspoken goal is mental healing. To reach that he wants to reconcile with his former enemies. A crucial point is therefore not only to find an audience for his traumatic experiences but also to find an audience that believes and trusts him. It is quite interesting but *My Heart of Darkness* does not mention any financial motives at all. The film gives the impression that the three black veterans join this boat trip with similar motives as the white South African. This parallels the idea of the communication around the campfire: symmetrical without any hierarchical structure. It is obvious that the film has difficulty maintaining this impression.

Already, in the opening scene of the film, the pictures from Maruis' time in Angola are central. Kept in a shoebox, at the end of the film they are burned in the cleansing ceremony conducted by a witchdoctor in the deepest bush. On the journey, Marius uses these pictures to validate his history. A discussion arises concerning the status of these documents. Patrick, Mario, and Sammy want to know where he fought in Angola, but Marius cannot answer these questions. He convinces them that he could not know because he was part of a special taskforce that was subjected to the utmost secrecy. To underline his argument, Marius shows the pictures. He is surprised by the reactions of the veterans. Patrick, as the spokesperson of the three, asks how it was possible to take pictures in a war while fighting

---

11 For information on the politics of culture in the GDR see: Jäger, Manfred (1994) Kultur und Politik in der DDR. Cologne: Edition Deutschland Archiv; or Emmerich, Wolfgang (1997) Kleine Literaturgeschichte der DDR. Leipzig: Gustav Kiepenheuer Verlag.

12 Precisely in this scene he tells that he is living by 600,- € a month and that he has to spent 140,- € of this for healthcare.

as a soldier: "If I go to the war and I am having a camera – that's not a war. [...] I fire, I never have a time to take pictures. [...] The thing is, I'm a soldier. If you bring a people dead from the other side. You can drop them at Walvis Bay. I can take a picture and that is my proof" (Julén/van Niekerk 2010:56.53). He then sums up his position: "For me, I don't trust pictures" (Julén/van Niekerk 2010:57.20). Van Niekerk understands photography as an index as Charles Peirce defines the term (cf. Wortmann 2007:174). In this approach the technical process of traditional photography (in contrast to digital) is seen as a safeguard for authenticity:

Sie speichert als Bild vielmehr die Totalität aller Ereignisse in dem Moment ihrer Aufnahme [...] Camera obscura und Fotografie gehören damit zu den analogen Medien, die 'im Unterschied zu Künsten [...] eben nicht darauf beschränkt [sind], mit dem Gitter des Symbolischen zu arbeiten. [...] Die Manifestationen von Ereignis und Welt als Bild apparativ generiert, entbehren jener Selektions- und Sinngebungsfilter, die jeder anthropologischen Äußerung vorgeschaltet sind – mit ihrer Absichtslosigkeit und Kontingenz stehen Fotografien sozusagen im 'Rauschen des Realen' – d.h. sie unterlaufen die Repräsentationsmodelle symbolischer und ikonischer Medien wie Schrift und manufakturelles Bild. (Wortmann 2007: 179)

This "advantage" that the physical and chemical processes stands for are of course double-edged. The result claims to be authentic and with this claim or promise they re-enter the world of discourse that they promised to transcend. Pictures as such do not know anything about the reality they display. They are not "der Welt entrissene Fundstücke [...] sondern Wirklichkeitsbilder, in denen wir unsere Erfahrungen, unsere Vorstellungen und Entwürfe von Welt verdinglicht als objektives Bild der Welt gespiegelt wiederfinden" (Wortmann 2007:167).

The confrontation between Marius and Patrick concerning the status of photography is followed by a short interlude showing the wild nature of the African countryside from the perspective of the boat. Van Niekerk's meditative voiceover dominates this sequence. He parallels nature to culture: "As we gather our scattered selves, nature too seems to reawaken. For decades, this bush was left in dead silence. Now slowly, timidly returning from exile the animal and the people. What is this wickedness in man set loose in such a place of beauty? Sacrificed in vain" (Julén/van Niekerk 2010:58.28). During the last words of this quote, we see a large crocodile gliding into the river. Surprisingly, the whole sequence is not shot in the "National Geographic mode". Some shots focus on the boat and on the doubting followers: Mario, Sammy, and Patrick are shown sleeping on the boat.[13] Their postures have to be described as stylized and not authentic at all. The sleep of the three veterans has to be understood as their exile from innocence and authenticity while negating the questions of guilt and reconciliation. This anacrusis is important for the following discussion. Marius complains about what he experiences as mistrust, saying that his history is seen as a fake. In quite aggressive words, he blames the others for mistrust. One can get the impression that he wants to force them to forgive him. Patrick sums it up in Portuguese: "He wants that we shall talk about that. Who wants to say anything?" (Julén/van Niekerk 2010:59.10). By overplaying the argument, one could say this: The white teacher is giving an exam to the black pupils and even if they were sleeping in class, now they will wake up. The discussion between the three connects with various aspects in the field of reconciliation studies, but it is difficult to see a common understanding. Nevertheless, Patrick in particular is learning very quickly and already knows how to please a disappointed teacher. He is in many ways the most interesting figure in the film. When he is introduced, the narrator emphasizes that he is a businessperson, and that he only accepted the invitaiton to join this expedition when he was told he could have his mobile phone with him. Playing on a slot machine, he is winning some money (Julén/van Niekerk 2010:24.00). This might be a negative reading of the film, but Patrick understands which goods Marius is asking (or paying) for and he delivers quickly. The discussion leads him to significant conclusions, and his point of view changes from day to night: "I think so what Marius tries to do, for me is very important, for me it's important. Very, very important ... Marius need that. I know we need that too. We gone help you to do that. I gone help you" (Julén/van Niekerk 2010:61.58).

This scene is not shot in the campfire setting but in full daylight, indicating the problematic status for van Niekerk. The authenticity of his documents are doubted and with this his project of regaining mental health/innocence. The three fellow travellers are in danger of losing touch with their notions of the archaic, natural, and mythic in van Niekerk's

---

13 Van Niekerk is not shown sleeping.

project. The turning point in the discussion is not anchored in the discourse, but comes from the outside. Van Niekerk suddenly observes a big crocodile and shouts that there is a "fucking huge crocodile" (Julén/Niekerk 2010:61.32) behind the boat. The crocodile that ended the voice-over sequence about the exile of man and nature breaks the surface of the river and the doubts of the veterans. The film's argument here is more metaphorical than discursive. After the re-entry of the crocodile, the chaos of the different opinions about reconciliation are streamlined into van Niekerk's position. The sleeping pupils have awakened.

Also in *Vaterlandsverräter* documents are of great importance and again the question of authenticity is at stake. In several sequences both Paul Gratzik and people which he observed are reading the reports he delivered to the Stasi. First Gratzik wants to reject Hendel's request and accuses her of attacking him in an unfair way. When he starts reading, his comments on the status of these documents are of great importance. First of all he denies being the author of the report that details the discussions with Heiner Müller. While reading, he recognizes that he is the author of the text, but the actual situation is still not clear : "O Gott, was für ein scheiss Deutsch [...] Kein Satz erinnert mich an die wirkliche Situation, so schlecht habe ich diesen Bericht geschrieben. Oder auf Band gesprochen – also wirklich eine saumäßige Leistung als Bericht" (Hendel 2011:32.02). He continues reading tearfully. As already mentioned, Heiner Müller was in some ways a mentor for Gratzik, and Gratzik saw in him a sort of elder brother. Just as in *My Heart of Darkness*, the setting for the realizations taking place in relation to these documents is very important. Gratzik sits alone in an anonymous hotel room looking over the city of Dresden. When he arrives at Dresden, he recalls the bombing of Dresden and his coquettish utterance that "he is going over corpses here" (see Hendel 2011:28.21) takes on another meaning. The reading of the report shows him that a reawakening of the dead is impossible and that reconciliation is out of reach. His betrayal of Heiner Müller is not something he can correct. That's why he reacts in such an emotional way, fighting back tears. Nevertheless, contrary to *My Heart of Darkness* – in which van Niekerk is also overwhelmed by his emotions (Julén/van Niekerk: 56.06) – this crisis is not resolved. Approaching the end of the report about Heiner Müller, Gratzik reaches a passage that is blackened out by the BStU[14] to preserve the rights of third parties. This is a very common occurrence when dealing with Stasi files today. Gratzik uses these blackened lines as a metaphor for his own personality: "Und jetzt ist es schwarz gestrichen und ich kann mich nicht erinnern. Irgendwas muss in mir sein, was ganz schlimme Sachen im Leben war, auslöscht also schwärzt. Na gut" (Hendel 2011:33.22). The subject is constituted by its lacunas and these gaps cannot be broached.

The reading-scene mirrors the four veterans sitting around the campfire but draws a very different conclusions. First, the status of the report is problematized. The terrible style is seen as a sign of the inauthentic. Some formulation shows Gratzik that he is the author of this report, but that are only glimpses without any significance for the whole. Furthermore, the document has gone through several stages of redaction and forms of media. Gratzik's utterance was recorded on tape in 1970 and the document is a transcript of the tape. While Marius manages to persuade his fellow travellers of the authenticity of his pictures, the documents in Gratzik's case remain inauthentic even for their creator. He recognizes his signature under the report, but the signature is not powerful enough to collect the divergent effects of his personality. Moreover, this signature is of course ambivalent as Paul Gratzik has signed with his code name: Peter. Gratzik is a person oscillating between Peter and Paul: Peter who disavowed Jesus Christ and Paul the man of letters. But there is no epiphany of truth and the viewer witnesses no transformation from Saul to Paul. While Saul overcame his blindness after the confrontation with the resurrected Jesus, the East German Paul is captured between insight and blindness. While in *My Heart of Darkness* the main theme is the return to origins, the common thread in Vaterlandsverräter can be seen in Gratzik's health problems. He is struggling with an eye-disease and the film team follows him on the way to his surgery. The one-eyed old man is in many ways symbolic for his unresolved situation and his lasting ambivalence toward the GDR and his involvement in that project.

---

14 BStU: The Stasi Records Agency is an upper-level federal agency of Germany that preserves and protects the archives and investigates the past actions of the former Stasi, which served as the secret police and foreign intelligence organization of the communist German Democratic Republic (East Germany). Since March 2011, Roland Jahn has been head of the agency. (wiki)

In a very telling way, Annekatrin Hendel reshapes this bundle of problems by using materials that are in contrast to common techniques in documentary movies. About a dozen times the interviewed persons are left behind and painted illustrations are used to show what they are saying. Gratzik sitting in a car on the way to Dresden depicts his breakthrough as a dramatist in 1968 with the piece *Malwa* and comments that this was a crucial point in his life. In his typical way, he presents this event as balancing on a knife edge: "In diesem Moment wurde mir bestätigt, dass, haben mir diese Menschen, diese hunderte Menschen bestätigt oder den Vogel eingegeben, dass ich etwas Besonderes bin" (Hendel 2011:17.35). It is uncertain if he is something special or if he only imagines this. Real greatness or idée fixe? The film illustrates this moment with paintings and a soundtrack of a radio playing that supports the significance of the paintings. We hear the buzz of the public at the premiere and hear the applause for the writer. The paintings that are used as illustration deepen the questions that are discussed here in a thrilling way.

At first glance, an anachronistic mood can be observed and understood as a marker of the past. The ravages of time are made visible. Furthermore, the style and color used in these paintings are clearly in debt to questions of genre. The sequence from Gratzik's breakthrough is closely connected to popular forms of the graphic novel and formula fiction. Hendel is here calling upon the melodrama, while in other sequences she picks up motives from spy movies and the romance novel. It is clear that she is playing with popular clichés in these parts of the film. Is she mocking her interview partner? It seems more likely that this strategy is being used in relation to the problem of the authentic.

First of all, the film shows no other traces of mockery.[15] At the end of the film, the director Annekatrin Hendel is shown in a discussion with Gratzik and she expresses that the experiment GDR has been a failure. She observes that Gratzik is again trapped in a circular argument about the GDR and himself – that the GDR has not been a failure. It is possible to read tenderness in the face of the director, as well as doubt, surprise and disbelief about this stubborn, old man, but we do not see irony. Secondly,

---

15 There is one exception. The picture of a Stasi spy sitting in a strandkorb intercepting some intellectuals surrounded by a huge crowd of people in an open-air bath is funny but the narration from the witness is too.

the use of this technique is not exclusively connected with Gratzik. When the reader of the West-German publishing house Rotbuch-Verlag is telling an absurd anecdote of her first meeting with the notorious writer Sascha Anderson in the lido of Pankow the same visual strategy is used. She also mentions this incident: "Aber jedenfalls war's so ein bisschen so ne Art Anti-Spionageplotte, die wir da miteinander aufgeführt haben, und ich fand das recht poetisch" (Hendel 2011:1.26.54). It is therefore more plausible to understand this strategy as problematizing the generic schemata that pre-structures the perception. If perception is governed by literary convention, the concept of authenticity is not innocent anymore. Paul de Man has described these problems in relation to autobiography:

"But are we so certain that autobiography depends on reference, as a photograph depends on its subject or a (realistic) picture on its model? We assume that life produces the autobiography as an act produces its consequences, but can we not suggest, with equal justice, that the autobiographical project may itself produce and determine the life and that whatever the writer does is in fact governed by the technical demands of self-portraiture and thus determined, in all its aspects, by the resources of the medium?".[16]

The autobiography is in the conventional view, as it is formulated by Lejeune, a genre of a more simple reference leading up to another contract of trust and believe between the recipient and the author.[17] Marius van Niekerk relies on this form of reference when he presents his photography and is shocked when their authenticity is doubted by Patrick and the other veterans. It can be argued that Patrick's critique is rooted in an even more rigid understanding of reference. He is most concerned to know where exactly Marius has been fighting, but his questions lead to a contextualization with political implications. Instead of this, Marius' project is first of all existential and he argues ad hominem with himself and his experiences as ultima ratio: "I know what I feel. I know how it feels in

---

16 De Man, Paul (1984:69) Autobiography as de-facement. In: De Man, Paul (1984) The Rhetoric of Romanticism. New York: Columbia University press.
17 Lejeune, Philippe (1975) Le pacte autobiographique. Paris: Édition du Seuil.

my heart" (Julén/Niekerk 2010:53,08).[18] Hendels re-mediations are pointing at the cracks in authenticity. In choosing these cliché pictures, she indicates that the perceptions of those involved and of the interpreter are governed by the needs of the media and the generic conventions – the people involved are framed by schemes that are outside of themselves. The authenticity and truth of a person is therefore an effect interwoven with other discourses. The boundaries between own and other are defied and the attributions of god and evil begin to circulate.

Vaterlandsverräter is not only showing Paul Gratzik and some of his friends from the cultural sphere of the GDR but also his case officer from the Stasi, Günter Wenzel. These people are often not very keen to talk in public so this is in some ways exceptional. Sitting at a table in a petty bourgeois kitchen, he explains how Gratzik was used as an informal collaborator (IM). Wenzel is not problematizing his own position in the surveillance of the people of the GDR at all. He is a representative of the instrumental rationality as it was criticized by Horkheimer in *Eclipse of reason* (Horkheimer 1947). He is giving a face to the thousands of Stasi agents who are characterized as narrow and uneducated people involved in pointless surveillance. To oversimplify a bit perhaps (and without equalizing the GDR with the Third Reich) Wenzel is an example of the banality of evil. The viewer is tempted to draw the conclusion that the case officer's relation to Gratzik is a cynical exploitation of the writer. This simplistic construction erodes when Hendel talks with Wenzel about Gratzik's roman *Transportpaule*. With this text, Gratzik came into conflict with the East German system of censorship[19] and several times the

---

18 This is a classical shift from object-authenticity to subject-authenticity that is critiqued by Andreas Huyssen. While the status of an object can be discussed and challenged it is much harder to challenge the emotional apparatus of an interlocutor. To mistrust the feelings of the other is also something almost forbidden and can devaluate the own position.

19 For this subject cf. Barck, Simone et.al. (1998) Jedes Buch ein Abenteuer. Zensursystem und literarische Öffentlichkeit in der DDR bis Ende der sechziger Jahre. Berlin: Akademie Verlag; and Westdickenberg, Michael (2004) Die «Diktatur des anständigen Buches». Das Zensursystem der DDR für belletristische Prosaliteratur in den sechziger Jahren. Wiesbaden: Harrasowitz Verlag

authorities denied him the permission to print. The cultural politics in the GDR in the years after 1976 were stormy because of the expatriation of Wolf Biermann.[20] In 1977 the novel was published, but the path to publication is stunning. Wenzel reveals that he was using his Stasi-internal contacts (to the division HA XX)[21] to get the novel through the eye of the needle. Hendel asks him: "Ist es wie ne Belohnung gewesen für die konspirative Arbeit?" And he answers: "Für den Paul? Na freilich! Die hätten den abblitzen lassen. Der hatte dort keine Lobby [...]" (Hendel 2011:1,06). The Stasi as publisher of literature that was critical to the system they should protect?[22] The writer exploiting the intelligence system to make a career? Undisputable answers are opened up again. Most obvious becomes this in the fact that the Stasi had records from Paul Gratzik as well as records on Paul Gratzik, both records as perpetrator and records as victim.

In contrast, the film *My Heart of Darkness* presents another model. While Hendel shows how reference and authenticity is merged in complex systems without a kernel, van Niekerk and Julén try to describe a journey to the beyond. The boat trip and the scenes around the campfire insinuate that it is possible to regain a lost innocence, to go beyond the suffered defilement of civilization. There is again a striking parallelism between the two films: *My Heart of Darkness* also uses drawings. When the historical background is described, a sketch of Africa is drawn and the involved parties are connected to the west and the east of Angola. This rather

---

20 Cf. Berbig, Roland red. (1994) In Sachen Biermann In: Protokolle, Berichte und Briefe. Berlin: CH. Links Verlag

21 "Die Hauptabteilung XX bildete den Kernbereich des Systems der politischen Repression und Überwachung des Ministeriums für Staatssicherheit. In Struktur und Tätigkeit passte sich die Abteilung mehrfach an die sich wandelnden Bedingungen der Herrschaftssicherung an." (Braun, Matthias (2008:3) Die Haptabteilung XX im Überblick In: Thomas Auerbach ed. et. al.: Anatomie der Staatssicherheit. Hauptabteilung XX: Staatsapparat, Blockparteien, Kirchen, Kultur, ‚Politischer Untergrund'. Berlin: Die Bundesbeauftragte für die Unterlagen des Staatssicherheitsdienstes der ehem. Dt. Demokratischen Republik, Abt. Bildung und Forschung)

22 Of course Wenzel used his influence neither to help friends or for esthetic reasons. The publication should tie Gratzik even more to the secret service.

childish sketch is made in strokes that remind one of charcoal. This is made explicit in later scenes when both van Niekerk and the other veterans draw sketches of battle scenes with charcoal. The metaphoric link to the campfire scene is obvious. Furthermore, the activity of drawing reflects the argumentation used in the film. Neither Marius nor his fellow travellers are educated artists. Their mode of sketching the awful battles reminds one of a child's drawings and supports the theme of innocence and purity. In the case of Samuel, the connotations to African art as primitive art are quite close. While Hendel uses the paintings as a crowbar to problematize authenticity and reference, Julén and van Niekerk use the drawings to highlight the sincerity of the creators. The paintings of children are imagined as products that are not subjugated to the laws of genre and therefore more authentic. They are not following the technical demands of a genre, and thus they are articulating the truth. The reference implies access to reality without detour.

Men in Boots (and Boats)

In the same way, Africa is imagined as a more authentic continent closer to the source of manhood and therefore less compromised. The travel up the river is clearly a journey to the origins to regain a rebirth free from all traumatic experience. In the opening scene, van Niekerk looks at the pictures from his time as a paratrooper and then stuffs them into a yellow box that he takes with him. The container is an ordinary shoebox with the imprint "Strassbergers Velskoene" – a South African brand. This package with his South African memories and the trauma they represent has to be delivered and of course destroyed. At the end of the film, the viewer witnesses a cleansing ceremony in the deepest bush, and all the threads of the natural, the authentic, and the original are bundled in this scene. The four veterans go through several stages and one is the killing of a scapegoat, after which their sins are forgiven. Now the package can be delivered to the fire. This part of the healing is shot at night and the whole village is gathered around the campfire. The four veterans commit to the flames symbols of their participation in the war and with that their violent memories. It is quite interesting that Mario for the first time uses the same vocabulary as Marius (Niekerk/Julén 2011:1.26.34). The pupils have again been keen to learn. This development is in many ways questionable. Van

Niekerk's project, as a personal project, is undoubtedly honorable and honest, but his politics of memory appear in these sequences as invasive, almost colonial. In several scenes, we see that his fellow travellers live in another world and that they are struggling with other problems. They are not white, middle class people, but they learn to behave this way in Marius' memory discourse.

This could be an interesting topic to deal with more intensively, but this paper wants to discuss the use of documents in relation to authentication processes. The film ends not with the cleansing ceremony and the burning of the pictures, but shows how the four veterans rebuild a palisade around Sammy's house that was destroyed in the night by a storm. The former enemies becomes co-workers, and it is not a surprise that the palisade forms a perfect circle. After this sequence the film breaks up the chronology and shows again van Niekerk arranging the now burned photos on the floor. By going back to the opening scene, the film suggests that the ring is closed – those involved can start again from scratch. This scene is also necessary because of an inherent paradox. Van Niekerk wished to destroy the pictures but in the filmic remediation, they get a new life. So his fear that his daughters will become aware of these documents is not resolved. This problem is commented on by the narrator, but he reformulates the goal in a slightly different way: "As children grow older there comes a day they find out about their parents history. I would like my two daughters to be proud of me and to know that I tried to fix the bad things that I have done during the war. And I'm sure that Patrick, Sammy, and Mario felt the same at the end of our journey, when we finally separated from each other" (Julén/van Niekerk 2010:1.31.41). The camera floats again slowly over the river into a beautiful dusk.

It's a strength of *My Heart of Darkness* that the film does not end with this tableau of reconciliation and rebirth. The last scene of the film shows Sammy and Patrick talking about the war as they exit the frame. They are laughing a lot. They confirm that war was a bad thing and then they fall into a discourse of "the good old times", reshaping battles with noises and gestures. This scene recalls the one where Patrick is telling the other veterans that he loved the war: "It was like a big party" (Julén/van Niekerk 2011:14.03). The circle of the film seems to remain open and it is possible to attribute this double ending to the divided directorship of the film. Van Niekerk gets his ending while Julén is opening up again to avoid the most colonialist implications. Nevertheless, the main plot stems from Van Niekerk's narration: the film is mostly dominated by his voiceover commentaries and his interaction with the other veterans. Already before the film reaches the double-end, there are cracks and contradictions in his concept of rebirth and reconciliation.[23] Through parts of the film, the idea of going back to the authentic, archaic, to get rid of memories, can be read as dream fulfilment: an illusion without any anchoring in reality. The strategies to create authentication are drawn into a maelstrom.

First of all, it has to be mentioned that the setting around the campfire is contaminated by a problematic issue. The notions of the archaic and the aura of direct communication has been mentioned (Urszene des Erzählens, Kayser). The film uses a lot of time to visualize these implications, but after a while, the viewer notices something that undermines these notions. In many scenes, Marius is looking at his counterpart in conversation and shows his compassion. This empathy is perhaps real but becomes hollow when the viewer realizes that he does not understand the language of the three other veterans.[24] This raises the problem of translation and simultaneously the problem of originality and transfer. Not only is it a practical problem (how precise are the translations?), but it also questions the possibility of direct communication undisturbed by interference. This is why the film under-communicates this problem and tries to create the impression of community in communication. This is a pattern that recurs with the figure that reflects these ideas most clearly: the witchdoctor. During the cleansing ceremony, the medicine man performs a form of litany, something the directors chose not to translate (there are no subtitles). In the figure of the shaman the notions of the archaic and the authentic are widened by the aspect of the sacred, the holy. He functions as an authority beyond the "transcendental homelessness". Divinity cannot be grasped by the human language and is at least untranslatable (cf. Niekerk/Julén 2011:1.24.44). The goal of the journey is rebirth through contact with the divine. The divine has to be the absolute other, not contaminated by human discourses. The film carefully builds up these notions, but in some sequences, there are elements that undermine and erode this concept. They

---

23 It is interesting that Skarpeid and Wagner observe a similar ambiguity in the last theme of the soundtrack (see: chapter seven in this book).
24 For the viewer, the problem is in some ways hidden due to the subtitles.

function like the punctum in Barthes well known essay on photography *Camera Lucida* (Barthes, 2000:26f.). The witchdoctor is shrouded in the skin of a leopard but his traditional outfit is dotted by two elements. First: As a shirt, he wears the Angolan flag. Second: He is wearing sneakers. The second element is not simply anachronistic but due to the metaphorical logic surrounding the shoes. The agent of the divine who will help van Niekerk get rid of his metaphorical "shoes of guilt" is infected by the Western culture and capitalism. Concepts of authenticity are here marked with contradictions and troubled.

It can be seen as coincidence, but in the setting with the campfire Marius is barefoot while his companions are wearing shoes. This argument is centered around the figure of Marius and his personal history so the equation of wearing shoes and guilt can not be transferred to the others. To go barefoot is a motif that emphasizes van Niekerk's journey as a penitent pilgrimage. In many religions, to go barefoot is a sign for penitence. The boat trip up the Kwando River becomes van Niekerk's own "walk to Canossa".[25] He gains absolution like the Holy Roman emperor Henry IV, the remorseful sinner, and in an almost uncanny way, Gratzik is excluded from this metaphorical discourse. Also in Hendel's film shoes play a role. Coming to Gratzik's farm a cold winter day the director brings hiking boots and the author comments upon them: "Und das sind die berühmten Schuhe. Donnerwetter. Junge, Junge." (Hendel 2011:4.50) Of course there is no direct link between shoes and guilt in Vaterlandsverräter, but these coincidences haunt the films and can push us close to over-interpretation.

But back on track. There are more recognizable patterns troubling the notions of absolution, for example, in the opening scene when Marius is in a hotel room arranging the photos and stuffing them into the shoebox. This scene is remarkable in that it also contributes to the discussion of authenticity and originality. The title of the film already references Joseph Conrad's classic novel. In adding the possessive pronoun "my" the tension between ownership and the foreign other is indicated. The boat trip is of course modeled on the novel, but the pretext is not stressed in a direct way. For example, there are no quotes from Conrad's text in the film. In some ways it is not surprising that the role of intertextuality is underplayed. Van

---

25 Cf. DeMello, Margo (2009). Feet and Footwear: A Cultural Encyclopedia. Macmillan. pp. 30–32.

Niekerk's project is governed by the desire to reach a zero point where he can start again. If the origin is contaminated with other concepts it is no zero point. Like the sneakers of the witchdoctor, intertextuality has to be suppressed to guarantee authenticity. At this point, the film gets entangled in some contradictions. The title of the film is indeed only a peripheral citation, but if we follow the track of intertextuality and adaptation from Conrad's seminal work another pretext is heavily present, that is, of course, Francis Ford Coppola's *Apocalypse Now* from 1979. The setting in the beginning of the film (hotel room, pictures, and the use of the rotating fan) are similar in both films. In the cleansing scene, the film reaches one of its peaks, marked by the killing of the scapegoat. The slaughter of an animal is also central in the delirious climax of Captain Willard's catabasis. The differences are of course obvious and profound, and it can be argued that *My heart of Darkness* counters Coppola's film. But that is not the point here. In this argumentation, it is more important to point out the technics the film uses to avoid turmoil. In some way it seems impossible to evade the wake/undertow from these cultural monuments, but the film tries to convey the impression that it is free from these influences. On the one hand, van Niekerk and Julén borrow authority[26] from seminal works of art in the Western tradition, and on the other hand, they have to repress the implications that occur through the play of this intertextuality.

Because of these concepts, van Niekerk's and Julén's film is in many sequences very stylized and attempts to reach an artistic entirety. While van Niekerk delivers his package and can travel back, for Paul Gratzik, the boat trip ends in another telling way. Landing with the boat in a little bay, Gratzik is confronted by bathers. He asks for help getting the boat higher onto shore. He thanks a young man and the following conversation evolves: "[Gratzik] 'Du hast nen deutschen Dichter eben an Land gezogen.' [Young man] 'Einen deutschen Dichter?' [Gratzik]: 'Juuh' [Young man]: 'Aber dann sind sie ja noch nicht tot.' [Gratzik]: 'Bitte?' [Young man]: 'Die meisten deutschen Dichter sind ja tot. Alle die bekannt und wichtig sind.' [Gratzik]: 'Ja siehste. Aber es gibt noch welche, die haben durchgehalten. Dazu gehör ich'" (Hendel 2011: 1.34.25). Hendel's film is open to contingency, and the absurdity of the situation cannot be translated

---

26 It is revealing to see that Trilling's work about authenticity has a chapter devoted to Conrad's Heart of Darkness.

into a unifying concept. In the same way that she exposes problems of authenticity in the use of documents, she doesn't cut the contingent elements: a naked man in this scene.

East German Coda

In her film *Anderson* (2014), Annekatrin Hendel experiments further with the tension between the authentic and the documentary. Also in this case she deals with her East German background and portrays the notorious Sascha Anderson. The paths of Gratzik and Anderson have crossed several times and there are striking parallels between these two writers. Due to his collaboration with the Stasi, Anderson became the undisputed star of the East German underground – the king of the Prenzlauer Berg. His motivations and destiny has been subject to much speculation. He wrote an autobiography[27] that disappointed the critics and has otherwise been quiet. He is still an enigmatic figure. To break his armor/shell, Hendel and her team rebuilt the kitchen of Ekkehard Maaß in a studio. Due to the illegal status of the underground scene in the GDR, they were forced to meet in private – mostly at Ekkehard Maaß. At first Anderson is shocked entering this room unarmed, but Hendel's purpose – to create a reenactment – is not successful. Anderson does not explain himself and the viewer is left in limbo: Is Anderson unwilling or not able to give an explanation? This is like the two lines from his poem that are used in the film as a leitmotif: "Vor dem Gartenhaus stehen drei Birken, die heißen Schuld und Sühne, ich weiß, welche die Liebste mir ist." But the last word is given to Paul Gratzik which named Sascha Anderson as his son: „Und dann is ja schwer heutzutage in der deutschen Sprache einen Satz aufzuschreiben den es vorher nich gab. Ja? Und das ist dem Sascha Anderson mehrmals gelungen … zwei oder drei Mal, glaub ich" (1:25:35).

---

27 Anderson Sascha (2002) Sascha Anderson. Cologne: DuMont Verlag

REFERENCES

Filmcredits

My Heart of Darkness – Sweden/Germany, 2010, 93 mins, directed by Staffan Julén and Marius van Niekerk. Coproduction of Gebrüder Beetz Filmproduktion, Eden Films, SVT and ZDF/arte. Supported by the Swedish Film Institute and Angolan ministry for Culture.

Flake – Germany, 2011, 44 mins, directed by Annekatrin Hendel. Produced by Itworksmedien.

Vaterlandsverräter – Germany, 2011, 90 mins, directed by Annekatrin Hendel. Corproduction of ARTE and Itworksmedien.

Anderson – Germany, 2014, 91 mins, directed by Annekatrin Hendel. Coproduction of Hessicher Rundfunk, Rundfunk Berlin Brandenburg and It works! Supported by Beauftragter der Bundesregierung für Angelegenheiten der Kultur und der Medien (BKM), DEFA Stiftung, Filmförderungsanstalt, K13 Kinomischung, Kulturelle Filmförderung Mecklenburg-Vorpommern and Medienboard Berlin-Brandenburg.

Literature

Anderson, Sascha (2002) *Sascha Anderson*. Cologne: DuMont Verlag.
Bachtin, Michail (2008) *Chronotopos*. Frankfurt a.M.: Suhrkamp Taschenbuch Wissenschaft.
Barck, Simone et.al. (1998) *Jedes Buch ein Abenteuer. Zensursystem und literarische Öffentlichkeit in der DDR bis Ende der sechziger Jahre*. Berlin: Akademie Verlag.
Barthes, Roland (2000) *Camera Lucida. Reflections on Photography*. London: Vintage books.
Berbig, Roland red. (1994) *In Sachen Biermann* In: *Protokolle, Berichte und Briefe*. Berlin: CH. Links Verlag.
Braun, Matthias (2008) "Die Hauptabteilung XX im Überblick" In: Thomas Auerbach ed. et. al.: *Anatomie der Staatssicherheit. Hauptabteilung XX: Staatsapparat, Blockparteien, Kirchen, Kultur, ‚Politischer Untergrund'*. Berlin: Die Bundesbeauftragte für die Unterlagen des Staatssicherheitsdienstes der ehem. Dt. Demokratischen Republik, Abt. Bildung und Forschung.

Daur, Uta red. (2014) *Authentizität und Wiederholung. Künstlerische und kulturelle Manifestationen eines Paradoxes*. Bielefeld: Transcript Verlag.

De Man, Paul (1984) "Autobiography as de-facement". In: De Man, Paul (1984) *The Rhetoric of Romanticism*. New York: Columbia University Press.

DeMello, Margo (2009). *Feet and Footwear: A Cultural Encyclopedia*. Santa Barbara: Greenwood Press.

Derrida, Jacques (^{4}1992): *Grammatologie*. Frankfurt a.M.: Suhrkamp Taschenbuch Wissenschaft.

Emmerich, Wolfgang (1997) *Kleine Literaturgeschichte der DDR*. Leipzig: Gustav Kiepenheuer Verlag.

Fischer-Lichte, Erika (^{2}2007:9-28) "Theatraliät und Inszenierung". In: Erika Fischer Lichte ed. et al. (^{2}2007) *Inszenierung von Authentizität*. Tübingen: Francke Verlag.

Huyssen, Andreas (2006:232-247) "Zur Authentizität von Ruinen: Zerfallsprodukte der Moderne". In: Knaller/Müller eds. (2006) *Authentizität*. München: Wilhelm Fink Verlag

Jäger, Manfred (1994) *Kultur und Politik in der DDR*. Cologne: Edition Deutschland Archiv.

Knaller, Susanne / Müller, Harro (2006:7-16) "Einleitung. Authentizität und kein Ende". In: Knaller/Müller eds. (2006) *Authentizität*. München: Wilhelm Fink Verlag.

Knaller, Susanne (2006:17-35) "Genealogie des ästhetischen Authentizitätsbegriffs". In: Knaller/Müller eds. (2006) *Authentizität*. München: Wilhelm Fink Verlag

Lejeune, Philippe (1975) *Le pacte autobiographique*. Paris: Édition du Seuil

Trilling, Lionel (1971) *Sincerity and Authenticity*. Cambridge MA: Harvard University Press.

Westdickenberg, Michael (2004) *Die «Diktatur des anständigen Buches». Das Zensursystem der DDR für belletristische Prosaliteratur in den sechziger Jahren*. Wiesbaden: Harrasowitz Verlag.

Wortmann, Volker (2006: 163-184) "Was wissen Bilder schon über die Welt, die sie bedeuten sollen? Sieben Anmerkungen zur Ikonographie des Authentischen". In: Knaller/Müller eds. (2006) *Authentizität*. München: Wilhelm Fink Verlag.

Memory, Trauma and Empathy[1]
On the (Un)representability of the Civil War in Art

NADINE SIEGERT

"There is no denying – we are a very wounded people. These works capture not only the horrors of war as portrayed by the mutilated bodies – the unimaginable damage to our souls, and the horrible violation of the common consciousness that emanates through every image."[2]
BRIGITTE MABANDLA/DEPUTY MINISTER OF ARTS, CULTURE, SCIENCE AND TECHNOLOGY SOUTH AFRICA

"In the autopsy I performed on my innocence I discovered the beginnings of secrecy and denial that must accompany the large-scale lie."[3]
COLIN RICHARDS

---

1 This title is inspired by Jill Bennetts seminal book Jill Bennett, Empathic Vision (Stanford University Press, 2005).
2 Fernando Alvim, "Memorias – Intimas – Marcas/Memory – Intimacy – Traces," Social Identities 5, no. 4 (December 1, 1999): 364.
3 Ibid., 276.

Angolan history is mainly a history of wars. After the long anti-colonial fight (1961 – 1974) that ended with the independence of the country in 1975, a civil war between the former independence movements started. In the 1980s, Angola was the venue for a proxy war between the Socialist bloc and the Western countries, supporting the different protagonists fighting on Angolan territory. After cold war ideologies crumbled in 1989, the war continued as a struggle for political hegemony and access to natural resources until 2002, devastating the land, economy, and social structures. These historical events undeniably led to widespread traumata. Direct or indirect violence has unfortunately become an enduring experience for most of the Angolan population. Many of Angola's peoples were scattered throughout the country or forced into exile making it difficult to continue cultural traditions. With Angola holding one of the highest rates of landmine injuries per capita in the world, amputees are a common sight in the streets of Luanda. Despite these hardships, the city might not be considered a *city of trauma*. But during the war, the city's infrastructure was fatally neglected, particularly in the informal neighbourhoods. Traumatic experiences of war were thus brought with the migrants and refugees to the city and are now situated within the biographies of the victims and perpetrators. Nevertheless, there is no public discourse on the civil war and no reconciliation policy comparable to that found in Ruanda, Sierra Leone or South Africa. This might be due to its adjacency, suggesting that traumatising events need several years before being far enough removed to be discussed. It produces a national amnesia, something that might also be politically motivated. In addition, in the field of cultural production, we face a situation where the recent past is to a great extent silenced, particularly the years of the civil war. Therefore, approaches to this issue by contemporary artists are probably even more worth examining, since they offer an engagement with a traumatic past that has been neglected elsewhere.

COLLECTIVE MEMORY AND THE ARTS

In the 1990s a few critical artworks dealing with history such as *Hidden Pages, Stolen Bodies* by António Ole and *Memórias Íntimas Marcas* by Fernando Alvim have emerged, but still the civil war is rarely to be found in contemporary arts today. These work are engaged with the collective memory, haunted by the phantoms of Portuguese colonialism, Portuguese fascism, the anti-colonial war, a civil war, socialism, and more recently, the emergence of a new capitalist economy that has made the country one of the leading economic powers of the continent. As members of a cosmopolitan elite, many artists have settled permanently or temporarily in the capital Luanda after the end of the war, having fled military service or having spent time pursuing education in Europe. After the ideological impact of the socialist cultural policy in the 1980s, the art scene has been rather silent on the local level, particularly in the 1990s when the civil war had its last heyday. Nevertheless, Angolan artists have been visible on the international level through their participation in the international Biennials of Johannesburg, Havana, and Dakar. The post-war situation, however, has provided space for utopian dreams of Luanda as an epicentre of contemporary art on the African continent, primarily fuelled by the curatorial strategies of Fernando Alvim in the context of the Luanda Triennial. In this context, interest in the history of colonialism and postcolonial conflict and its archives has also become a driving force in art production.[4]

In the following I will critically explore the perspectives employed by the artists dealing with these topics. My central question is if art might not offer *another*, alternative perspective on the engagement with the memory of the civil war that goes beyond the normative truth and reconciliation discourse dominant in southern Africa with its focus on forgiveness.[5] From

---

4 For more information on the Angolan art world see Nadine Siegert, "Luanda Lab – Nostalgia and Utopia in Aesthetic Practice," Critical Interventions 8, no. 2 (May 4, 2014): 176–200.

5 For an analysis of the reconciliation discourse in South Africa cf. Annelies Verdoolaege, Reconciliation Discourse: The Case of the Truth and Reconciliation Commission (John Benjamins Publishing, 2008); The relation of reconciliation and art is discussed in Stephanie Marlin-Curiel, "Truth and Consequences. Art in Response to the Truth and Reconciliation Commission," in Text & Image (Brunswick: Transaction Publishers, 2006); and Lizelle Bisschoff and Stefanie Van De Peer, Art and Trauma in Africa: Representations of Reconciliation in Music, Visual Arts, Literature and Film (London; New York: I B Tauris, 2012).

my understanding, this is also prominent in the film *My Heart of Darkness* discussed in this volume. In other words: Can art be a coherent form to remediate memory or does it rather point towards the un-representability of trauma? I will try to make sense of contemporary art practice and its relation to the history of the civil war, as invented and fictional as it might be.

Even if the comparison between a documentary film and works of contemporary fine art seems to be far-fetched, I think this juxtaposition allows some insights that are enriching for the debate unfolded in this volume. This is mainly because the artworks discussed in my paper and the film share some aspects in their motivation – to trigger a healing process on an individual and collective level. Using a close reading of the artworks, this paper wants to add an art studies perspective to the contributions dealing with memory in the context of the aftermath of violence. I will base my analysis of the artworks on theoretical frameworks that support the understanding of the relation between contemporary arts and memory studies: the notion of collective memory by Aleida Assmann and the critical engagement with the representability of memory and trauma by Jill Bennett.[6]

In particular, Assmanns notion of the *collective memory* is helpful to understand how activations and silencing work in cultural processes. According to Assmann, who follows the seminal writings of Maurice Halbwachs, collective memory is central for the cohesion of groups and as such forms a basis for collectivity.[7] She distinguishes between the *functional memory* as active part ("Funktionsgedächtnis") and the *storage memory* as passive part ("Speichergedächtnis"), the latter being the repository and background for latent memories. The "storage" is described as the total horizon of collected texts and images, a potential that can be activated and actualized from different perspectives.[8] Redundancy and loss are part of this process and create (dis)order in relation to historical and cultural preconditions. The notion of collective memory embraces both

---

6 Bennett, Empathic Vision.
7 Cf. Maurice Halbwachs, On Collective Memory (University of Chicago Press, 1992).
8 Jan Assmann and Tonio Hölscher, Kultur und Gedächtnis (Frankfurt: Suhrkamp, 1988), 13.

remembering *and* forgetting, both on an individual as well as a collective level.[9]

How is this understanding of cultural memory connected to artistic practice? Their engagement is not only inspired by a desire to excavate the hidden or hibernating stories but also to exorcise – metaphorically speaking – the demons in order to "repair" the past. These "demons" can be regarded as the haunting presence of the unresolved past that has to be released in order to allow new stories and images to become visible.[10] Relating this constructive potential with Assmann's thoughts on remembering and forgetting as the two central modes of collective memory, artistic work can entail a motivation to bring forgotten things back to memory – even if this means (re)inventing them – but also, conversely, to work towards the active process of forgetting by erasure or destruction. Thus, the artists discussed here work at the "frontier" between memory and forgetting, a shifting line that hides and reveals – and allows and denies – access to the cultural memory of the so-called border war.[11]

---

9   Aleida Assmann, "Archive Im Wandel Der Mediengeschichte," in Archivologie. Theorien des Archivs in Philosophie, Medien und Künsten, ed. Knut Ebeling and Stephan Günzel (Berlin: Kulturverlag Kadmos, 2009), 168; This is in line with Caruth's concept of the language of trauma as entangled in complex ways of knowing and not knowing. This language is played out not only in the repetitive haunting of memories, but also in representations of trauma in artworks. Cathy Caruth, Trauma (Baltimore, MD: JHU Press, 1995), 4.

10 On the discussion of "haunting" in cultural studies, see Avery F. Gordon, and Janice Radway, Ghostly Matters: Haunting and the Sociological Imagination (Minneapolis: University of Minnesota Press, 2008).

11 The notion of "border war" was mainly used in the South African context, where the border was not a demarcation of a state territory but an imagined frontier that was drawn against the perceived threat of communism. On the side of the communist MPLA in power, the war was communicated as a prolongation of the liberation war, a continuous fight against imperialism. In my article, I mainly use the notion of "civil war", which is more commonly used in Angola. On this discussion, see Gary Baines and Peter Vale, Beyond the Border War: New Perspectives on Southern Africa's Late-Cold War Conflicts, Revised. (South Africa: Unisa Pr, 2008).

The (Un)representability of Violence

The issue of adequate representation is a key question in the debate about art in relation to "events at the limits"[12], as the Angolan civil war certainly was. This problem is closely connected to the ethical responsibility in the visualisation of violence and the "pain of the others"[13], as Susan Sontag has formulated it in her seminal essay on war photography. How can artists differentiate from a sheer aesthetisation of violence? The fascination with the violence of war in arts had its heyday during the futurist movement in the early twentieth century[14], but more recently, with Adorno's pointed question about the role of art after the Holocaust[15], there has been no art making that celebrates war and violence without being questioned. The Frankfurt school philosopher formulated his cultural pessimism by questioning if it is morally acceptable to write poetry in the aftermath of Auschwitz. He later accentuated his argument and said that the representation of suffering is ambivalent: both unethical and necessary. Adorno didn't criticise the act of representation itself but the aesthetic pleasure that often results from it. From his perspective, art has the obligation to commemorate, but specific forms of representation are morally wrong.[16] Both Adorno and Sontag demand an active engagement with the topics of war and violence that goes beyond a mere consumption of images.

Considering the demands of Adorno and Sontag, we can argue that if a work of art that is both emotionally touching and rationally challenging enables a self-reflective process, an empathic confrontation with memory

---

12 Wendy Morris, "Art and Aftermath in Memórias Íntimas Marcas: Constructing Memory, Admitting Responsibility," in Beyond the Border War. New Perspectives on Southern Africa's Late-Cold War Conflicts, ed. Gary Baines and Peter Vale (Pretoria: Unisa Press, 2008), 158.

13 Susan Sontag, Regarding the Pain of Others, Reprint edition (New York: Farrar, Straus and Giroux, 2003).

14 Goedart Palm, "Das Format des Unfasslichen," Kunstforum International 165 (2003): 69.

15 Theodor W. Adorno, Noten zur Literatur, II (Frankfurt am Main: Suhrkamp, 1962).

16 Ibid.

that goes beyond aesthetisation or banalisation becomes possible. This is in line with the art theorist Jill Bennett's elaborations of the role of art in understanding trauma and loss. For her, the difference between an active and passive attitude towards the exposure of violence and the possibility of experiencing the trauma of another in a post-memorizing way is central[17]:

> How, then, might contemporary art engage trauma in a way that respects and contributes to its politics? If trauma enters the representational arena as an expression of personal experience, it is always vulnerable to appropriation, to reduction, and to mimicry. Is it possible, then, to conceive of the art of trauma and conflict as something other than the deposit of primary experience (which remains "owned" and unshareable even once it is communicated)?[18]

Bennett argues that art, which is able to actively engage with trauma, is not only communicative but also *transactive*, pointing towards relation and not representation. This works, because the artists work with encountered signs: symbols that are *felt* instead of *recognized* (in the sense of Deleuze[19]). The meaning of a symbol matters less than the question of *how* it works, its capability to trigger empathy or even an *"emphatic unsettlement"*.[20] Bennett defines this notion of empathy as a combination of emotional and intellectual operations, a deeply felt dismay that is able to change the perception or allows a critical engagement with an issue. In that sense, art should not translate or repeat the traumatic events but search for

---

17 Liese van der Watt has also developed a similar concept, working with the idea of "performativity": Liese Van der Watt, "Witnessing Trauma in Post-Apartheid South Africa. The Question of Generational Responsibility," African Arts XXXVIII, no. 3 (2005): 26–39; See also Liese Van der Watt, "Do Bodies Matter?: Performance versus Performativity," Arte. 70 (2004): 3–10; On this question see also Okwui Enwezor, "Remembrance of Things Past: Memory and the Archive," in Democracy's Images: Photography and Visual Art After Apartheid, ed. Jan Lundström and Katarina Pierre (Umea: Bildmuseet, 1998).
18 Bennett, Empathic Vision, 6.
19 Gilles Deleuze, Proust and Signs (New York: G. Braziller, 1972), 32.
20 Cf. Nicholas Mirzoeff, Shannen Hill, and Kim Miller, "Invisible Again. Rwanda and Representaion after Genocide," Trauma and Representation in Africa XXXVII, no. 3 (2005): 36–91.

an active, empathic testimony.[21] In contemporary art, such "transcriptions" of the experience of violence are often not limited to individuals; rather, the artists try to connect to trauma in a more abstract sense as a collective experience that also impacts whole societies in their post-memory work. To be an empathic challenge, the artwork – according to Bennett – has to be embedded in a broader societal context.

[Visual art] does not offer us a privileged view of the inner subject; rather, by giving trauma an extension in space or lived place, it invites an awareness of different modes of inhabitation.[22]

Further, she says that in this regard, artworks dealing with memory or trauma can even become political in their intersubjective character:

By figuring memory in "trauma art" as lived and felt in relation to a whole series of interconnected events and political forces, rather than as embodied in an atomized subject, we are able to move trauma into a distinctive political framework.[23]

TRAUMA ART

Bennett's position suggests that artworks that deal with traumatic events become involved in a negotiation of how collective trauma is perceived and processed. But what is meant by "trauma" in relation to art? Can we speak about "trauma art" when we analyse the contemporary artworks dealing with the civil war in Angola? In cultural studies, trauma is described in close relation to memory as the inability to remember or articulate a violating event. The original academic concept of "trauma" is more narrow and linked to psychoanalysis – which is only rarely considered in contemporary artworks. Later these psychoanalytical conceptualisations gave rise to a growing interdisciplinary field of study known today as Trauma Studies, which reaches across disciplines and engages in investigations of

---

21 Bennett, Empathic Vision, 7ff.
22 Ibid., 12.
23 Ibid., 18.

memory, witnessing and reconciliation, and also the relationship of these processes to cultural production.[24]

Etymologically, *trauma* derives from the Greek τραύμα, meaning "wound". Sigmund Freud defined psycho-trauma as damage to the psyche that occurs as the result of a violating event.[25] In 1932, Freud's student Abram Kadiner listed symptoms that would lead to the new diagnosis "post-traumatic stress disorder" (PTSD), a term that is still used today to describe the mental disorder suffered by war veterans.[26] Some characteristics of PTSD that can appear for long periods after the traumatizing events are a heightened inclination towards violence caused by traumatic incidences, feelings of inner fragmentation, nightmares related to violating events, sudden aggression, problems of identification, and fragmented memories. It is this last aspect, fragmented memories, that contemporary artists mostly engage with, maybe because it is possible to apply it beyond the individual to a whole society. This goes hand in hand with the application of the notion of "trauma" in cultural studies – literature and memory studies in particular – that discusses trauma in all aspects of society. In the context of studies on post-colonialism, it touches mostly on the negotiation of colonial traumas. In addition to trauma experienced by an individual, authors like LaCapra and Caruth describe a collective traumatised state of being that afflicts groups, societies, and even nations as a whole.[27] This recent trend in social and cultural studies brings trauma

---

24 Cf. Caruth's definition of trauma as wound inflicted upon the mind: Caruth, Trauma.

25 Sigmund Freud, Jenseits des Lustprinzips (Leipzig: Internationaler psychoanalytischer Verlag, 1920).

26 Cf. Greg Goldberg and Craig Willse, "Losses and Returns: The Soldier in Trauma," in The Affective Turn – Theorizing the Social, ed. Patricia Tincineto Clough and Jean Halley (Durham N.C.: Duke Univ. Press, 2007).

27 See amongst others Birgit Haehnel and Melanie Ulz (eds.), Slavery in Art and Literature (Berlin: Frank & Timme, 2009); Lisa Saltzman and Eric M. Rosenberg, Trauma and Visuality in Modernity (Lebanon, NH: UPNE, 2006); Shannen Hill and Kim Miller (eds.), "Trauma and Representation in Africa," African Arts XXXVIII, no. 3 (Autumn 2005): 1–4; Susan L. Jarosi, Art & Trauma since 1950, a Holographic Model (Durham: Duke University, 2005); Ana Douglass and Thomas A. Vogler, Witness and Memory: The Discourse of

closer to Assmann's concept of collective memory. In this sense, it becomes a collective loss of memory or a zone of collective amnesia within the cultural archive.

Thus in art, the trauma generated by violent events leads to a non-representability, such that it is impossible to represent and thus understand the cruelties that have been experienced by others. The non-representability on one hand, and the urge to deal with the horrors of the past on the other, both pose a complex dilemma for artists who engage in the field of memory and trauma. Nevertheless, many artists and writers try to visualise and write about traumatic experiences.[28] As we have seen, critics such as Bennett also concede art the ability to work with trauma in a positive sense. Historian Dominick LaCapra who, like Bennett, sees a capacity for healing in contemporary art, also supports this view, and he specifies this capacity as a form of "acting out" and "working through" the trauma.[29]

Considering these explorations of scholars engaged with the questions of the representability of violent memory and trauma, we can come to a preliminary conclusion that art has the ability to enable a secondary witnessing, an empathic experience of a muted trauma, something Hirsch calls a post-memory. In the following, I discuss a small selection of artworks dealing with the memory and trauma of the civil war in Angola and ask if these come closer to the form of "trauma art" demanded by LaCapra and Bennett than van Niekerk's filmic proposal of individual redemption and amnesia.

Trauma (London: Routledge, 2003); Richard Cándida Smith, Art and the Performance of Memory (Routledge, 2002); Dominick LaCapra, Representing the Holocaust: History, Theory, Trauma (Ithaca: Cornell University Press, 1996); Michael S. Roth, The Ironist's Cage: Memory, Trauma, and the Construction of History (Columbia University Press, 1995); Caruth, Trauma; Shoshana Felman and Dori Laub, Testimony: Crises of Witnessing in Literature, Psychoanalysis, and History (New York, NY: Routledge, 1992); Elaine Scarry, The Body in Pain (Oxford University Press US, 1985).

28 One of the first exhibitions showing a broad overview of "trauma art" has been: Gerald Matt and Angela Stief (eds.), Traum und Trauma. Werke aus der Sammlung Dakis Joannou, Athen (Wien: Hatje Cantz, 2007).

29 Dominick LaCapra, "Trauma, Absence, Loss," Critical Inquiry 25, no. 4 (1999): 700.

ARTISTS ENGAGING WITH THE PAST

In the film *My Heart of Darkness*, the box of photographs is a central motive. In the pictures we see van Niekerk's representation of the cruelties of the civil war. He did not shy away from a drastic representation of dead bodies as the epitomic image of war as discussed by Susan Sontag. The pictures were shot and then stored by the protagonist, thus becoming part of a deactivated cultural memory. But it only needs the protagonist to open the box to activate them again, and van Niekerk in the beginning of the film does precisely this. The narrative confirms the general assumption that the veteran's stories need to be told for the sake of individual healing and societal wellbeing.[30] Bennett however argues that in order to work with memory artistically, a simple representation might not be helpful in overcoming a certain traumatisation in the context of events such as a civil war. Is contemporary art providing a different approach?

One of the first works that deals with Angolan history from a contemporary arts perspective is certainly António Ole's *Hidden Pages, Stolen Bodies* (2001). This multimedia installation is an example that proves the artist's critical engagement with Angola's past. But even though this work was produced during the civil war, it doesn't address this conflict, instead it addresses the atrocities of late Portuguese colonialism and forced labour in particular. Ole's argument for the excavation of the hidden archive and the visualisation of unresolved collective memory has been paralleled by some stronger demands for a radical engagement with the past: Jo Ractliffe's photographic journeys into the landscapes of the aftermath of the civil war and Fernando Alvim's collaborative art project *Memórias Íntimas Marcas*.

Absence

Before focusing on a very relevant project by Angolan artist Fernando Alvim, I will first introduce another South African perspective that offers a different approach to the topic. The photographer Jo Ractliffe works on the "revelation of absence".[31] In Angola, she has realized the black and white

---

30 Baines and Vale, Beyond the Border War, 9.
31 http://artthrob.co.za/99mar/artbio.htm

photographic series *As Terras no Fim do Mundo*, where she also travelled to the sites of conflict. Like van Niekerk, she visited the site of the battle of Cassinga. Here, in 1978, a controversial air strike of the SADF on a refugee camp of the Namibian Liberation movement on Angolan territory (part of the *Operation Reindeer*) caused a high number of civilian casualties. For her project, Ractliffe accompanied veterans of the *61 Mech* battalion on their annual "Angola expedition" back to the battle venues.[32] This form of "travelling back" – probably with very different motivations – seems to be a regular veteran activity, and is not such a unique project as van Niekerk's film might suggest. For the veterans, these journeys are a form of reconciliation with their own past, whereas Ractliffe was motivated to translate her experience into images – mostly showing the absence of concrete traces of memory. She primarily found emptiness and an apparent invisibility of war. Nevertheless, she focused on this absence of obvious traces of violence in these abandoned spaces. Her project became an engagement with the aftermath as a moment when the landscape becomes – metaphorically speaking – an archive for ghostly memories or even traumas as unresolved memories.[33] Assmann's idea of storage memory as passive repository and background for latent memories is also applicable here. Landscapes can evoke memory but also point to its absence. The artist's engagement with these sites of the aftermath can be regarded as a form of memory work that is fundamentally different than, for example, monuments or objects of official commemoration. Travelling and working in situ can reveal the coexistence of different traces of memory in the traumatized landscape. The trauma is not extricable from the site, even if it is invisible.[34]

The photo series can be understood as a form of approaching the traumatized landscape of the Angolan civil war from the perspective of someone who was not actively involved but, as a sister and girlfriend, was impacted. Similarly to van Niekerk, Ractliffe works with mediated

---

32 The 61 Mech (61 Mechanised Battailon Group) is a veteran organization. Cf. http://www.61mech.org.za/

33 Cf. Okwui Enwezor, "Exodus of the Dogs," Nka. Journal of Contemporary African Art 25 (Winter 2009): 60–95.

34 Bennett speaks about a "double tracking" in this context. Bennett, Empathic vision, 98.

memories given to her by soldiers. Like him, she departs on a journey into strangeness, into a country that is not known to her but still matters to her understanding of her own history. She doesn't want to reconstruct this history, rather she tries to explore the conditions of her own perception through the generation of topographies of the present time in a landscape of absence. In her work, it is foremost the silence and emptiness, the lack of outspoken signs, that is disturbing. We *feel* that there is something under the surface. Her work raises questions instead of documenting the past violence.[35]

Whereas Ractliffe leaves the beholder troubled with this experience of the haunting presence of the unknown, Fernando Alvim's project goes a step further and tries to dig up this haunting presence. In order to find answers to the poignant questions, the three artists in his piece are actually digging in the ground. Alvim approaches collective memory from a psychoanalytical perspective, which brings him very close to the original understanding of trauma. In his work, trauma is understood as a collective loss of memory in a traumatized society and the inability to visualize and translate the experienced event. In this perspective, trauma has an ontological presence and means the absence of a collective processing of traumatizing events. The amnesia generated by this denial causes significant gaps in the symbolic order of a society.

Autopsy

Fernando Alvim has dealt with the civil war and the traumatic results related to Angolan history in a number of art works. One important aspect, that makes *Memórias Íntimas Marcas* an affective art project is to be found in Alvim's methodology. In this project, his conceptual framework was – even if sometimes semantically blurred – formulated quite clearly. His different works are often iconographically related to each other and thus form a group around metaphors from different semantic fields such as medicine and psychology (autopsy, psychoanalysis), religion (exorcism, catharsis), memory culture (archive, amnesia) and war culture (camouflage, sniper). Recurring motifs from the visual archive are flags, political and revolutionary icons, the figure of Christ, angels and fetish figures (*nkissi*)

---

35 Ibid., 85–86.

from the cultural context of the Bakongo.[36] In particular, in the works made in the 1990s, Alvim reflects on the possibility of re-membering the fragments of society left by a dystopian war situation.

Memórias Íntimas Marcas is the frame concept for a number of collaborative interventions organized by Alvim together with different international artists over a period of five years. The main point of departure was a twelve day process-oriented residency at Cuíto Cuanavale[37], one of the major battlefields where Cuban and South African military forces fought in the early 1980s. This first part was conceptualised in a kind of quasi-religious framework as a mythical journey and cleansing ritual.[38] In this context, he is not only artistically but also rhetorically developing the idea of an autopsy of the collective archive.[39]

MIM was organized as a collective psychodrama, an experience of exorcism, in order to emerge from the trauma of the Angolan war.[40]

Part of the project was a translation of the traces of memory and the engagement with the impact of the historical trauma into three-dimensional objects, the later exhibited at a number of shows in Angola, South Africa, and also Europe.[41] This was probably the first time that an artistic response

---

36 On this period in Alvims oeuvre see Simon Njami, "Fernando Alvim: The Psychoanalysis of the World," in Looking Both Ways. Art from the Contemporary African Diaspora (New York: Museum for African Art, 2003), 42–49.

37 Rory Bester, "Tracing a War," Nka. Journal of Contemporary African Art 9 (1998): Rory Bester says that this travel was 21 days long.

38 Cf. also Brandstetters analysis of the artwork as artistic fieldwork in search of lost memories: Anna-Maria Brandstetter, "Gewalt – Trauma – Erinnerung," in Entangled. Annäherungen an Zeitgenössische Künstler in Afrika, ed. Marjorie Jongbloed (Hannover: Volkswagen Stiftung, 2006), 122–55.

39 Cf. Fernando Alvim and Catherina Goffeau (eds.), Autopsia & Desarquivos (Brüssel: Sussuta Boé, 1999); Anne Pontegine, "'Memorias Intimas Marcas' – Brief Article," Art Forum International 38, no. 9 (May 2000).

40 Njami, "Fernando Alvim: The Psychoanalysis of the World," 45.

41 On the project see Alvim and Goffeau, Autopsia & Desarquivos. Alvim organized six exhibitions (including Luanda, Cape Town, Johannesburg and

to the Angolan civil war was shown publicly.[42] The project has gained some interest in academic discourse around art in relation to memory studies[43], but as an exhibition project, it is only sparsely documented and discussed. Morris describes that in the context of the exhibitions in South Africa, the show opened up a possibility to remember, to discuss, and to admit responsibility, including for former veterans who were partly engaged in the project. But the art project's first results were not so much redemption, but rather a possibility to "touch" the images that had been denied and neglected before. Those who partook in the project were willing to enter a space of collective mourning and to open up their wounds.[44] Rory Bester's critique of the exhibitions points to the unreflective framework of the violent archive the project is embedded in. He asks for a more conscious contextualization and careful application of terms such as exorcism and amnesia and a clear distinction between them. Bester argues that especially the collective archive, with its amnesic parts, has not been thoroughly conceptualized by Alvim.[45] I agree with Bester's critique when it comes to the politics of representation in an exhibition context where the question of the aesthetisation of violence has to be interrogated carefully, in

---

Pretoria and later also Lisbon and Antwerp), a book project (Autopsia & Desarquivos), a series of catalogues (Marcas News) and two films (Gele Uanga: War and Art of Elsewhere; Zinganheca Kutzinga: Blending Emotions) out of this initial project.

42 Morris, "Art and Aftermath in Memórias Íntimas Marcas: Constructing Memory, Admitting Responsibility," 159.

43 Cf. Bennett, Empathic Vision; Marlin-Curiel, "Truth and Consequences. Art in Response to the Truth and Reconciliation Commission"; Anna-Maria Brandstetter, "Gewalt – Trauma – Erinnerung," in Entangled. Annäherungen an Zeitgenössische Künstler in Afrika, ed. Marjorie Jongbloed (Hannover: VolkswagenStiftung, 2006), 122–55; Morris, "Art and Aftermath in Memórias Íntimas Marcas: Constructing Memory, Admitting Responsibility."

44 Cf. the individual statements by Colin Richards, "Aftermath: Value and Violence in Contemporary South African Art," in Antinomies of Art and Culture: Modernity, Postmodernity, Contemporaneity, ed. Terry Smith, Okwui Enwezor, and Nancy Condee, 2008, 250–89; Colin Richards, "The Names That Escaped Us," in Marcas News Europe, 4 (Antwerpen: Sussuta Boé, 2000).

45 Bester, "Tracing a War."

particular in the complex postcolonial space of the cold war.[46] But I want to focus on the first part of the project that, from my point of view, is clearly motivated by a conviction that artistic practise is a potent tool, able to intervene in the historical strata. This part of the project is not yet concerned with the methods of representation, which have rightfully been criticised by Bester.

For the first part of the project – the residency at Cuíto Cuanavale – Alvim invited other artists to return to southern African military conflicts, namely Carlos Garacoia from Cuba and Gavin Younge from South Africa. This selection is very meaningful in regard to the context of the proxy war situation in the 1980s. The title of the project refers to central aspects in regard to artistic work dealing with trauma: memórias = recollection; íntimas = inside; marcas = tracks or wounds. The project was built around the concept of the traumatized landscapes that form the archive of the civil war. In this understanding, a space can be regarded as carrier of memory that refers to an invisible past.[47] In this project, Alvim's artistic approach is similar to Jo Ractliffe's. In both Ractliffe's and Alvim's approaches, the artistic strategy does not involve the documentation of a spectacular moment but rather a search for traces. In this context, the artist becomes a witness who recognizes the neglected events in collective memory.

In *Memórias Íntimas Marcas*, the central intervention into that landscape archive is an artistic fieldwork at *Cuíto Cuanavale*, site of one of the most intense battles in 1987.[48] When the artists visit this place nine years later, the scenery is still marked by the conflict. Bullet holes in the houses are evident, but also abandoned armoured vehicles and, above all, the presence of landmines is obvious. On some house walls soldiers left graffiti describing their emotional state of mind during their deployment at *Cuíto Cuanavale*. Alvim's artistic motivation is directed towards a

---

46 On this debate cf. Wolfgang Muchitsch, Does War Belong in Museums?: The Representation of Violence in Exhibitions (Bielefeld: Transcript, 2013).

47 Narmala Halstead and Heather Horst, Landscapes of Violence: Unmaking, Forgetting and Erasing the Other (Blackwell, 2008); Bettina Fraisl and Monika Stromberger, Stadt und Trauma: Annäherungen, Konzepte, Analysen (Würzburg: Königshausen & Neumann, 2004).

48 On the role of Cuíto Cuanavale see e.g. Karl Maier, Angola: Promises and Lies (London: Serif, 1996), 33.

collective healing process, using this space as the setting for an engagement of the three artists with the place and the inhabitants. The artist himself describes this process as a form of *psychoanalysis*, a *cleansing* of the traumatized landscape and even an *exorcism*. Here, Alvim borrows Freudian terms to develop his approach as an engagement with the collective archive inscribed into the landscape. To enable a reconciliation or catharsis, it was necessary for him to extract the images of the war from that collective archive in a form of "autopsy". The idea that an open contact with the collective archive allows healing processes guides the artist's projects in situ as he works through the fragmented memories in the haunted space.

The singular projects developed in the context of the larger collective artwork had an affective rather than representative motivation. *Memórias Íntimas Marcas* offers an example of an approach that deals with trauma beyond the (im)possibility of its representation. The engagement with the landscape as site of memory turns it into an accessible archive that enables intervention. Different artistic approaches were developed during the residency period. In particular, the given landscape as well as the leftovers of human presence, such as the dilapidated houses covered with rough graffiti and bullet holes, were integrated into the performative actions. Cleaning and working through the sites was a major method in the artist's engagement. For example the artists washed the houses in a symbolic healing process. They also dug holes into the soil and pulled a two-headed doll through an underground tunnel in the soil in a kind of metaphorical birth – or, more aptly, a "miscarriage of death".[49]

The artist's choices were radical, but this *autopsy* of the haunting images might be the right choice to confront traumatic experiences, since the project enabled a dialogue with members of former hostile nations. It also leads to the possibility of an empathic and performative approach to an archive. Where contingency and trauma meet, performance – or art more generally – activates memory, official and otherwise, opening it up for the healing process and for the all-important task of re-working in the present. Here, questions of affect are key: what art *does* rather than what it *means*. Through the art of recalling, Bennett writes, an emotion becomes

---

49 Morris, "Art and Aftermath in Memórias Íntimas Marcas: Constructing Memory, Admitting Responsibility," 171.

perceivable.[50] Memory is figured as lived and felt rather than represented as meaning. This impact of the artistic process was most obvious in the reaction of the local inhabitants at the site of the former battle. After the performative engagement of the artists with the dolls, their symbolic rebirth in the underground tunnel, the villagers asked for one of the dolls to stay as a kind of spiritual device and they installed it on a ritual altar. This symbolic appropriation of the artwork underlines the power of the project and its performative and transformative potential.[51] The artists acted as the channels of this memory work for an as-yet mostly neglected and silenced event in the collective memory.

Mutilation

One example of the physical quality enacted in the project is Garacoia's excessive seven days of digging in the soil at Cuito Cuanavale. This might have been motivated by the need to lay bare the hidden memories and make visible what had happened, even if it was not there anymore, to dig up answers to his questions about the reasons behind this war.[52] But it was also motivated by the need to create a new territory that has been worked through with effort. He created visible pits and ditches, opening the metaphorical scars of the wounded territory again. Whereas the explosion of a mine had seriously injured his local guide, Garacoia himself was not wounded. Later, the video taping of the digging process became part of Garacoia's installation *In the Summer Grass*[53]. Gavin Younge was busy with cycling in the landscape of Cuíto Cuanavale, a form of remapping the landscape with a rather quotidian means of transport and always in danger of triggering a hidden mine. He later used the filmed material in his installation "Forces Favourites" at the exhibitions. One of Alvim's approaches was also a kind of mapping of the landscape using a camera

---

50 Bennett, Empathic Vision, 7.
51 Christian Hanussek, "Memórias Íntimas Marcas. Interview mit dem Künstler Fernando Alvim über den Aufbau eines Afrikanischen Kunstnetzwerkes", vol. x, 2 (Springerin: Hefte für Gegenwartskunst, 2004).
52 Morris, "Art and Aftermath in Memórias Íntimas Marcas: Constructing Memory, Admitting Responsibility," 163.
53 Brandstetter, "Gewalt – Trauma – Erinnerung," 142.

attached to a radio-controlled car. Parts of the footage were later used in the film *Zinganheca Kutzinga*. Perhaps taking the perspective of a crawling soldier hiding before an attack, this was a form of re-enactment of the characteristic imagery of warfare. After the residency, a number of works developed there have been transformed into multimedia installations and other artists where invited to join the exhibitions. In particular, the installations by the three artists who visited Cuíto Cuanavale speak the same language of empathy that was so important during their journey. Curator Thomas Miessgang describes the objects as "haptic, tactile phantasmagorias of an existence born out of loss"[54]. It is exactly this physical aspect that is most interesting for me in the discussion of the possibility of art as "working through" and "acting out" of trauma. For example, South African artists Jan van der Merve and Colin Richard showed installations with quasi souvenir objects from war such as cartridges or helmets. These objects are charged in a certain way as having been part of the actual events; the artists use them in an auratic sense. Carlos Garaicoa's contribution was a detailed graphic work drawn straight onto the outdoor walls of the galleries with an accompanying photograph showing a fist holding a whip touching a skull. Its title was "instrument to dissolve memory".[55] Alvim, on the contrary, loaded his objects differently. Puppets were suspended in the springs of a hospital bed (*El Hombre Solo* 1997) and another hospital bed carried a number of open scissors instead, all symbols not immediately associated with war. In his work *Difumbe* (1997-98), the Christ figure and the puppets float in glass boxes filled with water which emanate an unhomely presence by holding the puppets in suspension. The way Alvim installs them creates a specific scariness that might even be stronger in its affect than the representation of objects from the war. His objects work like fetishes, they are equipped with a certain power to trigger a mental catharsis. The objects are not souvenirs from the actual war but are rather charged by the artist's conceptual framework. One of these symbols is the figure of Christ (*Etranger* 1996-2000), but he is detached from the cross he is usually nailed to in religious iconography.

---

54 Miessgang, "Die Wunde des Seins. Neue Afrikanische Kriege und ihre Spiegelungen in der Kunst," 266–67.
55 Jan Wells, "Three Exhibitions Yorkshire Sculpture Park Wakefield," UnDo.net, January 7, 2005, http://1995-2015.undo.net/it/mostra/26678.

Christ becomes a floating signifier unable to promise redemption anymore but still carrying his religious meaning like a shadow. In 1995, during the first Johannesburg Biennial, Alvim also showed works related to the Angolan civil war, such as a prosthetic leg with an exposed femur and a burnt stump with the inscription "can anyone find my body". In the open space as well as in an installation called "Interventions" in the Angolan pavilion, Alvim showed sculptures of corpses, crucifixes and body parts, using a mixture of sculpture and installations that might serve as a metaphorical representation of the Angolan civil war. Whereas the leg sculpture was quite bluntly, the inside installation was more subtle and also integrated Alvim's semantically transformed words, a characteristic feature in many of his works.

Conclusion

As Terras no Fim do Mundo and *Memórias Íntimas Marcas* as Trauma Art?

Bennett states that in contemporary art, trauma is conceptualized as having a presence, a force.[56] This idea is very applicable to Ractliffe's and Alvim's approaches, both of whom are actually creating a counter-force through art in order to exorcise war-trauma.

One main difference between the motivation of the protagonist in the film and the artworks *As Terras no Fim do Mundo* and *Memórias Íntimas Marcas* is the dimension of the healing process that is triggered. Whereas Marius van Niekerk, the former South African paratrooper, seeks salvation for himself – epitomised in the quote "What does a man have to do to regain his self-worth after loosing it" – Alvim's art project aims to exorcise the collective trauma that not only paralyses Angola but also the former hostile countries South Africa and Cuba. This incommensurability can only be banned through aesthetic confrontation.

---

56 Bennett, Empathic Vision, 12; On the possibility to represent trauma in arts see also Aleida Assmann, Karoline Jeftic, and Friederike Wappler, Rendezvous mit dem Realen: Die Spur des Traumas in den Künsten, (Bielefeld: Transcript, 2014).

The idea is not only to remember and describe to the public details of this traumatic history, but, importantly, to educate, to attempt to exorcise its horrors, and ultimately to initiate a process of healing, both private and public, for those directly and indirectly involved. [...] presenting a 'common memorial' or an 'itinerary of intimate memories' generated by the personal experiences of the artists and their shared history.[57]

Whereas *My Heart of Darkness* lengthily shows the pictures of a mass grave at the battlefield of Cassinga, Ractliffe and Alvim avoid such drastic representations and work instead with metaphorical images and a physical engagement with the space as such. *Memórias Íntimas Marcas*, in particular, circles around the possibilities of healing a traumatised country, where this is meant both physically and metaphorically. The collective experience of travelling and being at a space that still carries marks of the battle's violence such as bullet holes and graffiti on the buildings, left over tanks and unexploded landmines, is comparable to both van Niekerk's and Ractliffe's motivation. The healing takes place through a collective endeavour, a return to the actual physical place. But Alvim embeds his project in the much broader framework of a national healing process that he coins with a term from the cultural field of religion (exorcism) on the one hand and psychoanalysis (amnesia) on the other.

Another aspect is the grade of participation and "experience" allowed. Marlin-Curiel points out, regarding different versions of theatre plays in the context of the TRC, that the ones really allowing societal healing were the ones that strengthened participation. This works when participants are "not watching an 'other' but are watching themselves. They are brought on a journey of healing as a community."[58] Van Niekerk lost his sanity to trauma, as he mentions in the film. To relieve himself of this burden – the haunting images from the past – he asks for the help and consent of three Angolans, but his approach is still very much about his own mental health. Similarly to Ractliffe and Alvim, the protagonist of the film feels the need

---

57 Alvim, "Memorias – Intimas – Marcas/Memory – Intimacy – Traces," 351.
58 Stephanie Marlin-Curiel, "The Long Road to Healing: From the TRC to TfD," Theatre Research International null, no. 03 (October 2002): 275–88; see also Marlin-Curiel, "Truth and Consequences. Art in Response to the 'Truth and Reconciliation Commission."

to get a picture of the cold war's inverted logic, the need to materialise this experience. Van Niekerk obviously speaks from his South African perspective, whereas the Angolans are addressed as a whole, neglecting the internal ideological fractions between the MPLA, UNITA and the FNLA, divisions that were the basis for the civil war conflict. The aspect of admitting responsibility is also probably present in both approaches. Whereas Van Niekerk departs on his frightening journey motivated by his need to understand the atrocities he himself has committed, he is also driven by guilt. He wants to come to terms with the shameful memories of the civil war that he has stuffed into a box and have haunted him ever since. He considers his own memories – in the form of photographs – as shameful and his main purpose is to get rid of them in order to avoid having his daughters ever witness them. The only witnesses allowed in this process are other veterans who, in his understanding, share a similar experience. His desire is to erase the past in order to make another life possible for the next generation. Even if he acknowledges the importance of going back to the places these memories are bound to and in working through them a second time, he probably does not reach the point of taking full responsibility for what he did, something that would mean accepting the inhumanity himself. Instead, he blames the war as the force that turns human beings into monsters. Nevertheless, integrating the perspective of the perpetrator into their work unites both Ractliffe's/Alvim's and van Niekerk's approach. They shift from a victim's perspective to the exploration of the possibility of forming narratives of power and attempts to critically intervene into the processes of memorialisation and cultural amnesia.

In my analysis of *As Terras no Fim do Mundo* and *Memórias Íntimas Marcas,* I focused on the aspects that make this work, in my understanding, an emphatic one that goes beyond representation. I see two important aspects that enable this: the involvement of the physical body of the participating artists (working through) and the choice of metaphorical rather than representational symbols. Why is this suffering important for the artists, who have never themselves undergone any the hardships of war and military service? It seems that all participants see it as a duty, an obligatory contribution of memory-work, for the healing processes of their respective collectivities. Some individual artist's responses clearly state the importance of actually having been at Cuíto Cuanavale during the aftermath, since this place has inscribed itself as mythical into the collective memory of all involved parties. Going there opens up the possibility of finding truth, as Garacoia points out in his film *Gele uanga*.[59]

Do the artists find ways to deal with the unrepresentability of collective memory when it is obscured by trauma or the legacy of the dominance of violence? The art works above might have shown that working with collective memory and trauma is an alternative way of dealing with the past and represents a new way of world making through emphatically engaging with the collective archive. The examples – *As Terras no Fim do Mundo* and *Memórias Íntimas Marcas* – are important as they open the hidden parts of the collective memory and write history anew in a sometimes very radical form. They deal with silencing and trauma by collectively and affectively exorcising the haunting past. Artistic practices that explore collective memory are motivated by a desire to heal and disturb at the same time. In this process, those who deploy such practices might be seen as imagining alternative futures as a means of transcending the trauma, something that becomes relevant through the trauma's excavation itself.[60] In different ways, all three artists address this process of mining the collective memory as means of rethinking and, ideally, reshaping the present and future.

Their motivation goes beyond representation and is not so concerned with the unrepresentability of history but rather proposes a process of *working through* the collective memory by affective associations.[61] In all the examples, memory and trauma – considered as incomplete and hidden, inscribed into the landscape or embedded in the collective memory of the civil war – are not taken for granted and their dominance is contested. Thus, the artworks also offer alternative readings of history by proposing ways to fill the gaps in memory and heal the trauma. By re-membering and dis-membering images of war atrocities, collective memory is stirred and disturbed and these works imply that a new order might become possible. This is more than just getting rid of a box of photographs by burning them. Harmonising is not the right strategy, but the engagement with traumatic

---

59 Marlin-Curiel, "Truth and Consequences. Art in Response to the Truth and Reconciliation Commission," 37.
60 Hal Foster, "An Archival Impulse," October. (2004): 22.
61 Cf. Ibid., 21.

memory must release something uncanny and inconvenient to keep memory alive and allow people to work through it, as LaCapra's points out with his concept of "emphatic unsettlement"[62]. In this conception, art is the opposite of amnesia due to its power to keep memory alive.

REFERENCES

Adorno, Theodor W. *Noten zur Literatur.* II. Frankfurt am Main: Suhrkamp, 1962.
Alvim, Fernando. "Memorias – Intimas – Marcas/Memory – Intimacy – Traces." *Social Identities* 5, no. 4 (December 1, 1999): 351–85.
Alvim, Fernando, and Catherina Goffeau (eds.). *Autopsia & Desarquivos.* Brüssel: Sussuta Boé, 1999.
Assmann, Aleida. "Archive im Wandel der Mediengeschichte." In *Archivologie. Theorien des Archivs in Philosophie, Medien und Künsten,* edited by Knut Ebeling and Stephan Günzel. Berlin: Kulturverlag Kadmos, 2009.
Assmann, Aleida, Karoline Jeftic and Friederike Wappler. *Rendezvous mit dem Realen: Die Spur des Traumas in den Künsten.* Bielefeld: Transcript, 2014.
Assmann, Jan, and Tonio Hölscher. *Kultur und Gedächtnis.* Frankfurt: Suhrkamp, 1988.
Baines, Gary, and Peter Vale. *Beyond the Border War: New Perspectives on Southern Africa's Late-Cold War Conflicts.* Revised. South Africa: Unisa Pr, 2008.
Bennett, Jill. *Empathic Vision: Affect, Trauma, and Contemporary Art.* Stanford Calif.: Stanford University Press, 2005.
Bester, Rory. "Tracing a War." *Nka. Journal of Contemporary African Art* 9 (1998): 64–66.
Bisschoff, Lizelle, and Stefanie Van De Peer. *Art and Trauma in Africa: Representations of Reconciliation in Music, Visual Arts, Literature and Film.* London; New York: I B Tauris, 2012.

---

62 LaCapra, "Trauma, Absence, Loss."

Brandstetter, Anna-Maria. "Gewalt – Trauma – Erinnerung." In *Entangled. Annäherungen an Zeitgenössische Künstler in Afrika*, edited by Marjorie Jongbloed, 122–55. Hannover: VolkswagenStiftung, 2006.

Caruth, Cathy. *Trauma*. Baltimore, MD: JHU Press, 1995.

Deleuze, Gilles. *Proust and Signs*. New York: G. Braziller, 1972.

Douglass, Ana, and Thomas A. Vogler. *Witness and Memory: The Discourse of Trauma*. London: Routledge, 2003.

Enwezor, Okwui. "Exodus of the Dogs." *Nka. Journal of Contemporary African Art* 25 (Winter 2009): 60-95.

- "Remembrance of Things Past: Memory and the Archive." In *Democracy's Images: Photography and Visual Art After Apartheid*, edited by Jan Lundström and Katarina Pierre. Umeå: Bildmuseet, 1998.

Felman, Shoshana, and Dori Laub. *Testimony: Crises of Witnessing in Literature, Psychoanalysis, and History*. New York, NY: Routledge, 1992.

Foster, Hal. "An Archival Impulse." *October* (2004): 3-22.

Fraisl, Bettina, and Monika Stromberger. *Stadt und Trauma: Annäherungen, Konzepte, Analysen*. Würzburg: Königshausen & Neumann, 2004.

Freud, Sigmund. *Jenseits des Lustprinzips*. Leipzig: Internationaler psychoanalytischer Verlag, 1920.

Goldberg, Greg, and Craig Willse. "Losses and Returns: The Soldier in Trauma." In *The Affective Turn – Theorizing the Social*, edited by Patricia Tincineto Clough and Jean Halley. Durham N.C.: Duke Univ. Press, 2007.

Gordon, Avery. *Ghostly Matters: Haunting and the Sociological Imagination*. University of Minnesota Press, 2008.

Haehnel, Birgit and Ulz, Melanie. *Slavery in Art and Literature*. Berlin: Frank & Timme, 2009.

Halbwachs, Maurice. *On Collective Memory*. University of Chicago Press, 1992.

Halstead, Narmala, and Heather Horst. *Landscapes of Violence: Unmaking, Forgetting and Erasing the Other*. Blackwell, 2008.

Hanussek, Christian. "Memórias Íntimas Marcas. Interview mit dem Künstler Fernando Alvim über den Aufbau eines Afrikanischen Kunstnetzwerkes". Vol. x. 2. *Springerin: Hefte für Gegenwartskunst*, 2004.

Hill, Shannen, and Kim Miller (eds.). "Trauma and Representation in Africa." *African Arts* XXXVIII, no. 3 (Autumn 2005): 1-4.
Jarosi, Susan L. *Art & Trauma since 1950, a Holographic Model*. Durham: Duke University, 2005.
LaCapra, Dominick. *Representing the Holocaust: History, Theory, Trauma*. Ithaca: Cornell University Press, 1996.
- "Trauma, Absence, Loss." Critical Inquiry 25, no. 4 (1999): 696-727.
Maier, Karl. Angola: *Promises and Lies*. London: Serif, 1996.
Marlin-Curiel, Stephanie. "The Long Road to Healing: From the TRC to TfD." *Theatre Research International* null, no. 03 (October 2002): 275-88.
- "Truth and Consequences. Art in Response to the Truth and Reconciliation Commission." In T*ext & Image*. Brunswick: Transaction Publishers, 2006.
Matt, Gerald, and Angela Stief (eds.). *Traum und Trauma. Werke aus der Sammlung Dakis Joannou*, Athen. Wien: Hatje Cantz, 2007.
Miessgang, Thomas. "Die Wunde des Seins. Neue Afrikanische Kriege und ihre Spiegelungen in der Kunst." In *Africa Screams. Das Böse in Kino, Kunst und Kult*, edited by Tobias Wendl, 263-70. Wuppertal: Peter Hammer, 2004.
Mirzoeff, Nicholas, Shannen Hill, and Kim Miller. "Invisible Again. Rwanda and Representaion after Genocide." *Trauma and Representation in Africa* XXXVII, no. 3 (2005): 36-91.
Morris, Wendy. "Art and Aftermath in Memórias Íntimas Marcas: Constructing Memory, Admitting Responsibility." In *Beyond the Border War. New Perspectives on Southern Africa's Late-Cold War Conflicts*, edited by Gary Baines and Peter Vale, 158-74. Pretoria: Unisa Press, 2008.
Muchitsch, Wolfgang. *Does War Belong in Museums?: The Representation of Violence in Exhibitions*. Bielefeld: Transcript, 2013.
Njami, Simon. "Fernando Alvim: The Psychoanalysis of the World." In *Looking Both Ways. Art from the Contemporary African Diaspora*, 42–49. New York: Museum for African Art, 2003.
Palm, Goedart. "Das Format des Unfasslichen." *Kunstforum International* 165 (2003): 64-97.
Pontegine, Anne. "'Memorias Intimas Marcas' – Brief Article." *Art Forum International* 38, no. 9 (May 2000).

Richards, Colin. "Aftermath: Value and Violence in Contemporary South African Art." In *Antinomies of Art and Culture: Modernity, Postmodernity, Contemporaneity*, edited by Terry Smith, Okwui Enwezor, and Nancy Condee, 250–89, 2008. "The Names That Escaped Us." In Marcas News Europe. 4. Antwerpen: Sussuta Boé, 2000.

Roth, Michael S. *The Ironist's Cage: Memory, Trauma, and the Construction of History*. Columbia University Press, 1995.

Saltzman, Lisa, and Eric M. Rosenberg. *Trauma and Visuality in Modernity*. Lebanon, NH: UPNE, 2006.

Scarry, Elaine. *The Body in Pain*. Oxford University Press US, 1985.

Siegert, Nadine. "Luanda Lab – Nostalgia and Utopia in Aesthetic Practice." *Critical Interventions* 8, no. 2 (May 4, 2014): 176-200.

Smith, Richard Cándida. *Art and the Performance of Memory*. London: Routledge, 2002.

Sontag, Susan. *Regarding the Pain of Others*. Reprint edition. New York: Farrar, Straus and Giroux, 2003.

Van der Watt, Liese. "Do Bodies Matter?: Performance versus Performativity." *Arte*. 70 (2004): 3-10.

"Witnessing Trauma in Post-Apartheid South Africa. The Question of Generational Responsibility." *African Arts* XXXVIII, no. 3 (2005): 26-39.

Verdoolaege, Annelies. *Reconciliation Discourse: The Case of the Truth and Reconciliation Commission*. Amsterdam: John Benjamins Publishing, 2008.

Wells, Jan. "Three Exhibitions Yorkshire Sculpture Park Wakefield." *UnDo.net*, January 7, 2005. http://1995-2015.undo.net/it/mostra/26678. Last access April 2017.

Miss Landmine in Angola

Negotiating the Political Aesthetics
of the Mutilated Body

NORA SIMONHJELL

My Heart of Darkness (2010) deals with the complex consequences of the Angolan war. During the documentary four men are confronted with their past history and actions as soldiers. They have a common past, but different memories and different ways of dealing with their choices during and after the war. The film editor, Marius van Niekerk, initiates this process. The documentary functions as a kind of memory work. In a war, a soldier takes part in an active way. He has to follow orders, and acts as a part of an ongoing conflict. However, a war is not only a struggle between soldiers. During the years of conflict the war affected the lives of civilians in many ways. It is estimated that millions of innocent people were killed or badly injured during those years of conflict.

Angola is often referred to as a "Landmine nightmare".[1] After several decades of civil war, landmines and landmine injuries have become a huge problem in Angola. Angola obtained its independence from Portugal in 1975. According to the UN there are millions of unexplored mines left in the ground. Angola signed the UN Landmine Ban in 1997, and has had an

---

1   According to the web site "Our Africa", over 70 000 persons are amputees as a result of landmine accidents in Angola. 8000 of them are children. This is an estimate from 2012. The process of removing landmines is very slow, and the number of people getting hurt increases every year. See http://www.our-africa.org/angola/land-mines, visited 02.01.2016

extensive demining program.[2] Despite this, landmines are still a huge danger to the population and particularly to the young. People, particularly women and children, often get injured by the remaining mines: women because they work in the fields, and children because they play in the same places. Because of this, Angola has a large population of disabled people. This creates huge problems both for the ones who have been injured, and for society as a whole. The lack of good and functional prostheses is also a large problem. The landmines are a reminder of the war, and, in a way, the war continues long after the conflict causing it has been solved. The injured persons' living conditions are often very challenging after their accidents. The landmine question draws our attention to civilian life, and it makes us aware that this is a gendered question. While most of the soldiers are men, many of the people hurt in landmine accidents are women and children. In this article, I am going to discuss an international art project that uses this duality to question the aftershock of the Angolan war from a gendered perspective, and deals with the consequences of the civil war from a very different perspective than Van Niekerk.

The Norwegian artist Morten Traavik (b. 1969) is globally known for his politically engaged art.[3] When he visited Angola, the huge number of landmine victims struck him. He wanted to find a form that could make the destinies of the people disabled by landmines part of the political, aesthetical agenda and that could also raise public awareness. He focused on young women and staged a beauty contest for disabled landmine victims

---

2 The UN has an intensive disarming program. Angola was one of the countries that signed the international anti-personnel Landmine Convention in Ottawa, 1997. See https://www.un.org/disarmament/convarms/landmines/, visited 02.12.2016.

3 Traavik's official web site is http://www.traavik.info/. Traavik is both an artist and a theatre instructor. He has an outspoken political agenda for his art and his projects always fall on the brink between events and official performances. His art is discussed worldwide and some of his works have provoked discussion not only as art, but also as political interventions. Traavik works in the international art scene, and most of his works are done outside Norway. He has provoked people with his art collaboration with North Korea. Traavik is inventive and uses his art to question political issues. This is also the case in the Miss Landmine project.

in 2008: *Miss Landmine*.[4] As the title *Miss Landmine* indicates, the women participating in this beauty contest were visibly affected by landmine accidents. The most striking thing about the contest was that the women's injured bodies were exposed with their scars, stumps, and missing limbs. Traavik addresses the reality of war from an outside position and through an aesthetic approach.

In *My Heart of Darkness*, Van Niekerk's outside position is quite different. He was a soldier during the war. He returned to Angola after several years to confront his past actions and his memories. This journey back forces the three other former soldiers (Angolan and colored) to be a part of his process. Unlike the other articles in this volume, the central message of *My Heart of Darkness* is not the focus here. Instead, I am going to focus on aspects that are under-communicated in the documentary. Traavik's art project *Miss Landmine* puts the lived experiences of dealing with war injuries and the disabled female body in the spotlight. The art project underscores that there are always innocent people getting hurt during times of war. The history of war is inscribed in their bodies.

The *Miss Landmine* contest in Angola 2008 was followed by the *Miss Landmine* contest in Cambodia the year after. Angola and Cambodia are ranked alongside Afghanistan as the world's most heavily mined places, with a correspondingly high number of victims suffering the terrible effects of the weapons left after years of conflict. I am not going to address the contest in Cambodia here. The structure of this event was very similar to the one in Angola. Later in this article I am going to discuss the general structure of beauty pageants. After the actual beauty contests, the *Miss Landmine* project was developed into a photo exhibition that travelled the world, and also exists as a web page. My background material for this article is mostly from the web page.[5] The photo exhibition has been

---

4   All the information about the Miss Landmine is from the art project's official web page. See http://miss-landmine.org/, visited 26.06.2016.

5   Miss Landmine is a multi-media and global art project. My article is mostly based on the art project Miss Landmine's web page. I have also studied the catalogue accompanying the Cambodian Miss Landmine project in 2009 (Traavik 2009). I have visited the art exhibition Miss Landmine in Stavanger, (09.14-23, 2012). During the literature festival, Kapitel, some of the photos from Miss Landmine were exhibited in fashion stores. For a full overview of the

launched in different galleries in Europe, and the project has received a great deal of international media interest.[6]

The *Miss Landmine* art project is an alternative way to spotlight ongoing landmine tragedies. Every day people get hurt and lose their limbs due to mine accidents. The title of the art project counterpoints the words "landmine" and "miss". It contrasts the serious and tragic reality of landmines with the glossy celebration inherent in a beauty pageant. *Miss Landmine* is also an example of a different way of posing a political question. It is an aesthetic investigation of the personal and social consequences of war. The exposition of the hurt, mutilated, and gendered body tells quite a different story of war from the talking soldiers in Van Niekerk's film. This is a story often silenced in war history. I am going to focus on the representation of the female disabled body and how this particular body was staged in a beauty queen setting. What kind of beauty contest was *Miss Landmine*? How does this pageant deviate from other beauty contests like *Miss World* and *Miss Universe*? And how does the exposing of the mutilated and amputated body challenge the standardization of beauty norms? What are the structural similarities and differences in such different beauty contests? I will be drawing upon disability theory and beauty queen theory.

LANDMINES AND BEAUTY

Let me start out with some facts about landmines and the campaigns against them. The International Committee to Ban Landmines and Jody Williams won the Nobel Peace Prize in 1997.[7] Today, at least ninety-nine countries are affected by mine areas, and thirteen countries still produce mines, including the US, Russia and China. A statistic says that here are

---

different staging of this art project see the official web page. There are also several reviews of the project there. All the photos in this art project are taken by the Norwegian photographer Gorm Gaare.

6   The Angolan Miss Landmine project was sponsored by the EU, the Norwegian Art council and the Angolan government's de-mining commission and a charity run by the country's first lady.

7   https://www.nobelprize.org/nobel_prizes/peace/laureates/1997/

almost 500,000 land mine survivors in the world. The campaign against landmines' greatest success came about in 1997 when the Ottawa Treaty, banning the production and use of landmines, was signed by many countries. Despite this important work and global awareness, landmines and landmine injuries are still a huge problem in many parts of the world. Earlier in the same year as the Ottawa Treaty, the late Princess Diana (1961-1997) brought the issue of landmines to global attention when she called for an international ban on landmines during a visit to Angola with the International Red Cross. She met several victims of landmine accidents and was deeply moved by their stories.[8] Princess Diana was known globally for her beauty and warm heart. She used this status as a princess to increase international awareness of people's living conditions.

Princess Diana's work putting the question of landmines and the brutal reality of landmine victims in the spotlight of media and public awareness is an important background for understanding the art project *Miss Landmine*. Princess Diana's beauty, social position, privileged life, and "perfect body" functioned as a distinct contrast to the social and embodied reality of the injured people in the project. Princess Diana's emotional involvement was seen as incredibly controversial because, at that time, the British army was still using landmines.[9] In Princess Diana's commitment, the grotesque reality of the landmine victims and the international concept of royal beauty were brought together for the first time. She showed an affective dedication that raised a lot of public attention. At a beauty pageant

---

8   Princess Diana visited Angola in the middle of January 1997. BBC 1 made a documentary on her travel: "Heart of the Matter" in February 1997. Princess Diana is the narrator in this documentary, and her voice reveals how shocked she was by the harsh reality of the landmine-injured women and children she met. Britain had landmines at that time, and the princess' trip raised a great deal of political discussion. The documentary is available on YouTube under the title "Princess Diana's landmine program" (p 1-3). At the end of the documentary, Princess Diana speaks: "Before I came to Angola, I knew the facts, but the reality was a shock". Later that year Princess Diana spoke on the issues of landmines at different events all over the world.

9   The United Kingdom ratified the Ottawa Treaty on July 31, 1998. See https://en.wikipedia.org/wiki/List_of_parties_to_the_Ottawa_Treaty, visited 26.06.2016.

the most beautiful woman is crowned as a semi-royal queen. This is also the case in the *Miss Landmine* contest. However, the mutilated bodies of the women in the contest make a huge visual statement and function as an affective reminder of the consequences of the use of landmines. Loosing legs and arms are the most common injuries of landmine accidents. People's bodies are transformed. The victims of landmine accidents face new challenges in life. They might face trauma, phantom pains due to lost limbs, and they might find it challenging to accept their new bodies and loss of function resulting from their injuries. As disabled women, they might face problems with daily activities, and some will have problems getting jobs and so on. This art project's staged pageant represents the possibility of escaping this reality for a moment, and is a way to underscore that the landmine-injured women are more than victims.

BEAUTY PAGEANTS AS FORMATTED EVENTS

Miss Landmine uses the same structure as international beauty contests like *Miss Universe* and *Miss World*. Every year, all over the world, young women try their luck in these contests. As the fairy tale saying goes: Mirror, mirror on the wall – who is the most beautiful of them all?

Beauty contests are both universal and diverse. In their introduction, the editors of *Beauty Queens on the Global Stage* argue that these contests "put gender norms – conventionally, idealized versions of femininity – on stage in a competition awarding the winner a 'royal' title and crown" (Cohen, Wilk, and Stoeltje 1996, 2). Beauty contests are places where different sorts of cultural meaning are produced, consumed, and rejected – where local and global, ethnic and national, national and international cultures and structures of power are engaged in their most trivial but vital aspects. The most striking aspect of a beauty contest is that the formula is so easily adapted from one cultural setting to another. The competitions are formatted, and the format is "easily reproduced and recognized" (Stoeltje 1996, 13). Let us take the *Miss World Contest* as a structural example.[10]

---

10 All the information on the Miss World Contests is from the web page www.missworld.com. It is worth mentioning that a woman from Angola was

There is a first round where local young women compete for a place in a national final, and then there is a final where the national beauty queen is crowned. This contest qualifies the winner for the international final. Then women from more than 100 countries gather in the international final. The final is a month long event, with several preliminary galas, dinner balls and different activities. They even compete in swimming suits and gowns. The final is broadcasted live and the field is narrowed down to 15-20 delegates before one of them is finally crowned *Miss World*.[11] The winner spends one year of travelling to represent the *Miss World Organization* and its various causes. This is a privileged position, and "The Queen is granted a voice to speak publicly, although she always speaks as a sign of the unit she represents" (Stoeltje 1996, 28). This was also the case for the late Princess Diana. It was her royal title and social position, including her mediated persona, that gave her the ability to visit Angola and speak on behalf of the landmine-injured women and children.

The women taking part in *Miss World* or *Miss Universe* competitions are chosen from huge numbers of women that in a sense "embody the best of a nation". Their outer beauty and bodily features are selected to represent the good morality and good intentions in the society they represent. The beauty contests are glamorous events and the participating women are all young women and "natural beauties". Their bodies have harmonious proportions, and they are tall and slim with long hair. They embody the symbolic ideal of the *innocent* and *pure* female – not married and not mothers. The question of "inner beauty" and morality is set up as equal to their outer beauty. "To talk of 'beauty'" is to emphasize physical features when pageants are supposed to find a representative for the community who embodies that local people believe to be the best of themselves: talent, friendliness, commitment to the community and its values, upward morality" (Lavenda 1996, 31). This makes a rather complex duality, and they are "dialogic hybrids", as Lavenda puts it (1996, 32). The women compete in beautiful dresses and bikinis, and the beauty contests often have a semi intellectual side – the candidates must formulate

---

crowned Miss Universe in 2011. This made her the first Angolan to win an international beauty pageant.
11 There are also other titles in Miss World such as Beach Beauty, Miss Talent, Miss Sports and others.

arguments for their causes.[12] The gowns worn at the final event contribute to an elegance that goes beyond all the outfits the contestants have worn previously.

Each description of a beauty contest includes encounters between different systems of knowledge and structures of power. They are produced and played out as public events. The beauty contests create an image of the female according to cultural beauty norms, and judge the contestants according to that norm. Here one has to consider woman as a cultural category, her body as a beauty norm, and the individual and particular woman: how her actual body matters. When international beauty contests emerged, these beauty norms became global. This has affected the global understanding of beauty.

Beauty contests are also "struggles over the power to control and contain meanings mapped on the bodies of the competitors" (Cohen, Wilk, and Stoeltje 1996, 9). The contest is a highly ritualized performance: "The ritual contest in which young women compete against each other for the title of queen, serves to link the individual contestants to the sponsoring institution as a representative of it" (Stoeltje 1996, 15). The beauty pageant defines the ritual contest of the beauty queen, creating a form that embodies the message and communicates it both to participants and audience. These two mechanisms are combined in the beauty contests in order to link together the individual female, specific units of the community and the community as a whole (Stoeltje 1996). The female participants' bodies might be understood as signs that embody symbolic values and represent an ideal. What about women whose bodies do not fit the narrow beauty standards represented in *Miss World* and *Miss Universe*? *Miss Landmine* is a subversive political art project that challenges the standardized expression of beauty.

The photos show three of the women taking part in the Angolan *Miss Landmine* pageant in 2008. Their bodies all have visible marks from the war. On the left is Miss Cuanza Sul. She is wearing a swimsuit. Miss Huíla is wearing a blue flowy dress, and Miss Mexico is wearing a green dress.

---

12 It is very easy to joke about beauty queens and beauty contests. The cultural stereotype is that the girls that take part in contests like this are blond and naïve. The joke goes: "I am against war and for peace".

Photo by Gorm K. Gaare, reprint by courtesy of the artist.

At first glance, they look almost like ordinary beauty queens. Two of them are dressed up in beautiful dresses, and one wears a swimming suit. They are pictured on the beach with a beautiful sunset, and they all have a ribbon that marks them out as special: contestants in a pageant. They are all posing for the camera. However, if we take a closer look, the differences are huge. The striking difference lies in their bodily features. One of them is pregnant, and they have all lost one leg in landmine accidents. The missing limbs are exposed, and so are their crutches and walking sticks. They are styled and posed just like beauty queens, but they all have amputated legs. The disabled and black woman bodies are visible political and embodied expressions. The most interesting aspect of *Miss Landmine* is that all the stereotypical understandings of beauty norms are challenged from within. The women taking part in the *Miss Landmine* project challenge a narrow understanding of beauty. Even though the landmine project is following the same structure as an average beauty pageant, the women's life stories and real life experiences and disabled bodies expand this frame. Their bodies have inscribed upon them the country's war narrative and brutal reality, and the women taking part in it represent a large diversity of female life experiences.

MISS LANDMINE AS A DEVIANT BEAUTY PAGEANT

Miss Landmine is a highly political work of art and an example of an artistic counter practice. First and foremost the female contestant's bodies are not neutral. Their visible scars and amputations bear witness to the results of war and show that the results of war go on among the civilian population long after a peace treaty. Women's life stories during war time are often silenced, and this art project highlights how the actual and personal body and life story are influenced and affected by a country's greater history. The contestants in Angola were all women of colour. Colour and gender are political since beauty contests started as competitions for white middle class women in Western Culture, while the *Miss Landmine* project is taking part in Africa and Asia and the contestants are relatively poor.

The pageant was staged according to the same model and structure as ordinary beauty pageants. In the actual competition, ten women representing their provinces met on stage. They all wore make up and beautiful dresses. The final was held in the capital of the country and the prize was a golden, state of the art prosthetic limb. The prosthetic limb was later replaced by a personal adapted prosthesis. The women were introduced with a picture, their name and age, what province they represented, their family status, and an explanation of how they were injured and what happened to them. In addition to this, their current occupation was listed as well as their favourite colour. Each of the contestants disclosed these details as part of the event. A final piece of information marked the unusual matter of their bodily appearances, their actual landmine accidents and their amputations.

Miss Landmine imitates the structure in global beauty pageants. The *Miss World* organization raises money for "good causes", money that mostly goes to children's charities. "Beauty with a Purpose" is the slogan, which was added to the contest in the 1980s. The *Miss World* candidates are judged by "intelligence and personality" as well as by their outer beauty. This represents a merging of inner and outer beauty. We can compare this to *Miss Universe*, where the slogan is "The Woman with the Stars". Traavik draws upon the *Miss World* slogan "beauty with a purpose", using "beauty with a difference" for the *Miss Landmine* project. The political side of the project lies in the ironic twist of the slogan. The project challenges concepts of female beauty, the notion of beauty contests, and most importantly how we look at disabled persons and especially disabled women. What is the difference? The most interesting part in Traavik's project is the reuse of a highly commercial genre and its transformation into a subversive aesthetic and political practice. The most interesting thing about *Miss Landmine* is that it is a staging of a beauty contest for women that does not correspond to the commercial beauty standards and norms. The artist's outspoken intention was to put the landmine problem on the international agenda, to show these women's situation, and to pose questions about femininity and the female body.[13] He wanted to raise these women's self esteem and to show that there is beauty in all people. At the photos and the final event they pose in both swimming suits and long gowns. Their prostheses and artificial limbs are exposed and visible – this is the most striking difference! The winner of *Miss Landmine* Angola 2008, was the thirty-one year old Augusta Hurica. She won the jury's vote. Maria Restino Manuel (26) won the internet competition.[14]

Beauty contests evoke controversy over what qualities should count in a competition, how woman should act, and what the outcome means. "At every level, beauty contests include a process of selection and representation, of making highly public choices that assert some kind of

---

13 When the Miss Landmine contest moved to Cambodia (2009) Traavik posted this manifesto on the project's web page. Manifesto – Miss Landmine: EVERYBODY HAS THE RIGHT TO BE BEAUTIFUL!
- Female pride and empowerment.
- Disabled pride and empowerment.
- Global and local landmine awareness and information.
- Challenge inferiority and/or guilt complexes that hinder creativity – historical, cultural, social, personal, Asian, European.
- Question established concepts of physical perfection.
- Challenge old and ingrown concepts of cultural cooperation.
- Celebrate true beauty.
- Replace the passive term 'Victim' with the active term 'Survivor'
- And have a good time for all involved while doing so!

http://miss-landmine.org/cambodia/index.php/manifesto.html

14 All the information about the candidates is based on the webpage Miss Landmine. See http://miss-landmine.org/

collective identity" (Wilk 1996, 230). As such they are always arenas for the definition of locality, for inclusion and exclusion. But when local choices are exhibited on a wider stage and become part of a hierarchical structure, the pageant becomes much more than this. We are forced into a discussion, implicit or explicit, of the hierarchical relationship between the local and the regional, the national and the international, and there different pageants can showcase "local standards of beauty" (Cohen 1996, 129). I would like to add that it is not only the global–local discussion that is staged in a beauty contest. These contests also address the physical and personal body as such. The different female bodies visible on stage are symbolic signs – they are both political and individual – thus they represent both the women individually and their connection with their nation and its difficult history.

On a fundamental level, beauty pageants are a spectacle – "a public performance that changes the relationship between people and culture" (Wilk 1996, 231). When a culture is objectified and performed on stage, then beauty is commodified and publicly contested. It shifts the audience away from participating in culture and towards treating objectified culture as another consumer culture – commercial mass culture. This also implies a standardization of the understanding of beauty. According to Moslalenko, "the standards of physical beauty are becoming standardized throughout the world, on many occasions against the will of the common population" (1996, 73). The problem of sending a coherent message is embedded in the structure of the beauty itself. The meaning of representations put on stage is always negotiated and contextualized in interactions between contestants and audience.

THE DISABLED BODY

In the media debate surrounding the *Miss Landmine* project, some critics have said that this project exposes the female landmine victims as freaks, and thus criticized the art project.[15] I am not going to go into that discussion, but there are some overlaps with Traavik's project and the idea

---

15 The reviews on the project have been diverse. For an overview see the project's web page.

of the sideshow. "People who are visually different have always provoked the imaginations of their fellow human beings" (Garland - Thomson 1996, 1). The freak is seen as someone different, often someone that has a deviant body. Freak shows were very popular in the late nineteenth century, the most famous one being the American P.T. Barnum's sideshows (Garland - Thomson 1996, 5), which featured people with severe disabilities as "stars". This way of exposing bodily differences is highly problematic (Bogdan 1988). Even though *Miss Landmine* is exposing women with their amputated limbs, scars, and stumps, I would argue that the art project is not a freak show. This is caused by the underlying ironic twist and play with the beauty pageant format and the understanding of "the miss". The "misses" in *Miss Landmine* expand the category in an accurate way.

Miss Landmine falls in between politics, art, and public debate. The art project is an aesthetical and thereby a cultural and political intervention in the conflict. *Miss Landmine* articulates both personal histories and the country's general history. The personal, local and national histories merge on the global art scene.[16] The women taking part in the competitions have visible marks on their bodies and have all experienced real trauma in life. The individual body matters – it is a visible, embodied, and lived representtation of the country's history.

Feminist and queer theory has drawn attention to the concept of the other. These theories have contributed to an interesting critique of the social, gendered, political and embodied complexities embedded in cultural and mediated signs. Disability theory theorizes disability in ways similar to how feminism has theorized gender. Garland Thomson (1997) states that the disabled body is extraordinary. She questions the representations of disabled bodies, and argues that in spite of the huge interest in "the other",

---

16 The global media interest in Traavik's project has been huge. Miss Landmine's official web page has published a selection of articles from all over the world. The artist has been interviewed on the radio and television and in newspapers from Norway to New Zealand. From my perspective, it is interesting to see the differences in the media representation. The Norwegian press has a tendency to underscore that Traavik is a Norwegian artist, and that he received funding for his project from both the Norwegian Foreign Department and the Art Council. The international press, it seems to me, is more interested in the thematic side of the project.

it is the physical other – the disabled person – who is still left in the dark or silenced. Although a lot of recent scholarship explores how differences and identity operate in such politicized constructions as gender, race and sexuality, according to disability theorists, cultural and literary criticism has generally overlooked the related perceptions of corporeal otherness, something we think of variously as "monstrosity", "mutilation", "deformation", a person who is "crippled" or simply as physical disability (Davis 2006, Davis 1995) (Siebers 2008). Identity, subjectivity, and the body are cultural constructs to discuss. All representation is political.

Disability theory and disability studies form a quickly growing interdisciplinary field that questions the representation of disability in media, popular culture, and history. The aim is to reframe "disability" as another culture-bound, physically-justified difference to consider along with race, gender, class, ethnicity, and sexuality. Concepts like "the cripple", "the invalid", and "the freak" are all questioned: One aim is to deconstruct the cultural encodings of these and other "extraordinary bodies". This means to question concepts and understandings of "the normal" and stereotypes, and to unravel the complexities of identity production within social narratives of bodily difference. A disabled body challenges our understanding of the body, and can, according to Garland-Thompson, be addressed as an *unruly body*. An unruly body challenges beauty standards, concepts of entity, harmony, and the dualism between ill and well. In disability theory, it is common to make a distinction between *impairment* – which is a physical condition like lacking an arm or a leg, or being paralyzed or blind – and *disability* – the social construction of that physical body. The impairment is the physical lack that limits one person's activity. By using concepts like "lack" and "limits", I have used stigmatizing language. Disability is still mostly seen as bodily inadequacy or a catastrophe to be compensated for with pity or good will, rather than accommodated by systematic changes based on civil rights. Reckoning with bodily variations is to address embodied otherness. Disability is the unorthodox made flesh, refusing to be normalized, neutralized or homogenized, as Garland-Thomsen puts it (1997, 2009). One might say that this is a classical construction of a "we"–"they" dualism. The disabled woman is the cultural opposite to the commercial beauty queen. The material form of the body put on stage is not neutral. But who is "we" and "they" in this dualism? The women in the *Miss Landmine* represent the double other: they are of colour and disabled.

Beauty contests put norms and values into question. Although contests like *Miss World* and *Miss Universe* intend to be non-political events, they raise highly interesting questions. The most obvious and striking pertains to gender norms and the concept of beauty. The winner is crowned and becomes thereby a kind of semi-royal person. An average girl is transformed into a highly valued queen like in the fairy tale Cinderella. She represents a form of idealized female beauty standard – and also an idealized version of the femininity and the female body. The crowning of the beauty queen is a ritual and a rite of passage which transforms the average girl into someone special (Stoeltje 1996). This is also the fact in the *Miss Landmine* project. But the crowning of *Miss Landmine* represents an opposition to the highly standardized beauty norms staged in *Miss World*. These women's bodies are special and bear the marks of a personal life story. Their mutilated bodies create new challenges for them – both on a personal and social level – and they have to find a way to live with disability. The pregnant woman's participation in the art project is highly interesting from this point of view. She stands in extraordinary opposition to the slim, thin, and "pure" girls in the global beauty shows.

Let me therefore underscore one more point about the disabled body. When a person has a visible disability, this might challenge our gaze. To stare is a form of power relationship. It is an asymmetrical form of power. Garland-Thomas argues that the stare is a form of intensified gaze (2009). The extraordinary bodies of the women in this pageant challenge idealized beauty norms. The contestants are all beautiful women, yet they have unruly bodies. Their mutilated bodies and cultural embodiment exposes how their individual fates are a result of their country's brutal history. To stare at them is also to look at an individual, cultural, social, and historical utterance of trauma. Their survival stories are dramatic narratives and those stories are told by their participation on stage. These are crucial elements in the exposed and embodied beauty – and this makes them extraordinary.

Focusing on a bodily feature to describe a character throws the spectator into a confrontation with the character that is predetermined by cultural notions about disability. It can be easy to pity a person with severe bodily impairments. What does this pity do? The *Miss Landmine* project exposes the *beauty of real women* – not idealized young girls – but real women marked by life itself. Some of them are married, mothers, pregnant, divorced or widows, and some of them are older than the average girls that are contestants in beauty pageants. They have all experienced trauma in their lives. Traavik tries to show them as living women, not as victims or disabled. The amputated limb tells a silent but highly dramatic story. All the women have survived dangerous accidents, and they all have to live their lives in a body that forever has a visible mark manifesting both their individual fate, but that also acts as an inscription of the history of the country in which they live. Their country's violent history is forever inscribed on their bodies. The physical body is important in itself; it becomes both an individual and personal site and bears witness to the traumatic events in the injured woman's lives. On a higher and more general level, it becomes a representation of the country's destiny – thus "the body is a primary site for the inscription of cultural truth" (Johnsson 1996, 99). The cultural truth in Angola is that a large number of the population has landmine injuries. This is a tragedy – but the women in *Miss Landmine* are all presented as strong and independent women.

VISIBILITY MATTERS

The women taking part in the *Miss Landmine* project might be labelled "the other" for many different reasons: They are of colour, disabled, and they represent the "third world". The structure of international beauty competitions is a conduit for proliferating Western styles, values, and expectations (Cohen, Wilk, and Stoeltje 1996). But these women also highlight contrasts among local and globalized notions of beauty and identity, and the local conditions which enable resistance against global styles. Themes of identity, citizenship, and affirmation come to the surface in these contests. "Beauty contests qualify unambiguously as local forms that also exist in a hierarchy linking the local to the global" (Stoeltje 1996, 18). The body on stage embodies a certain visibility – even though it challenges the same visibility – just as the women in *Miss Landmine* do. By inviting and staging women deviant from "normal" female bodies, Traavik's art project is highly political. The damaged, hurt, and amputated female body is often hidden, or made silent. By giving these women access to take part in the same beauty structure as the "normal" beauty queens, the project exposes their pain, life experiences, and disability. It also exposes how war's aftermath has consequences for both individuals and the society as a whole.

Feminist theorists have been critical to beauty contests. They represent idealized beauty images and not "real women". They impose dangerous notions of body image. Traditional beauty contests exclude women who do not fit into given beauty norms. They promote the illusion that there is in fact a beauty standard, that beauty can be measured objectively, and that beauty has a concrete existence apart from the individual. They define beauty as a norm or ideal standard, and imply an objectification of women (Craig 2002). By promoting a sense of consensus around beauty, these contests' narrow notions of diversity, reduce the range of possibilities for individual expression, and allow special interests and small constituencies to speak for the majority. Social, contextual, and subjective notions appear biological, universal, and absolute. There is always a division between front stage and backstage in a beauty contest: what is shown on stage and the meaning hidden behind it. There is always something more going on than the beauty contest itself. On the one side, the highly commercial event and the idealized beauty standards, and the idea of global and universal beauty. On the other side, we have the "misses": differences in background, social class, the personal, and the political. The big question is: what is shown and what is not shown?

"The standards of physical beauty are becoming standardized throughout the world, on many occasions against the will of the common population" (Moskalenko 1996, 73). Who decides what is beautiful? Alternative beauty contests have been arranged for different reasons. The most striking is the history of *The Black Beauty Queen*. Black women and women of colour were not allowed to take part in US beauty pageants, so communities of colour made their own show in the sixties (Craig 2002). There have also been alternative beauty contests for fat or plus size queens (Snider 2009). Fat studies developed from feminism and queer studies, and like disability studies, are growing in the US. Snider discusses beauty pageants for fat or obese woman. Snider argues that from a fat or queer perspective, the thin woman is seen as a part of the heteronormative domination. Heterosexual and thin is the norm, and fat and obese woman are traditionally excluded from conceptions of beauty.[17] The fat beauty pageant is structured and staged in part as a self-visualization strategy, and willingly plays with prejudices against the fat: the fat lady as lazy, silly, lacking self-control, and so on.

The list of alternative beauty queens could easily be enlarged. My point here is that Traavik's use of the beauty contest format as the structural framework of an art project stands in a subversive tradition. The commercial format is twisted and reused as a political statement. By putting the alternative, extraordinary, hidden, or in other ways unruly body – black, disabled, and injured – on stage, *Miss Landmine* becomes a way of visualizing other realities, other life stories, and other destinies, the real stories of other women. In doing this, the art project contributes to a larger understanding of the consequences of the country's brutal war history.

CLOSING REMARKS

Traavik's engagement in the Angolan history is both interesting and problematic. His outside position starts from a perspective very similar to a tourist's. His encounter with the Angolan war history and civil wars was triggered by the fact that many of the people he met during his stay in Angola had been hurt by landmines. This made him go further into the context and landmine issue. This led to aesthetic intention and aesthetic action which addressed political and historical issues. Even though the intentions behind this art project were good, it is important to remember that Traavik is a white, male artist, and comes from one of the richest countries in the world. His outside position is defined by gender, economic difference, and questions of class, and there is clearly a huge contrast between the artist behind the project and the women taking part in it. This aspect should have been discussed even further. My intention in this article has been to address the art project's potential as an aesthetic way of raising political awareness about the personal and embodied issues of the landmine-mutilated female body.

---

17 See for example the discussion on beauty pageants for transvestites in the Philippines (Johnsson 1996).

References

Our Africa. Angola http://www.our-africa.org/angola/land-mines1.2016
Miss Landmine: http://miss-landmine.org/, visited 02.01.2016
UNODA. United Nation Office for Dissarment Affairs: https://www.un.org/disarmament/convarms/landmines/, visited 02.12. 2016
1997. Heart of the Matter. https://www.youtube.com/watch?v=HJO0bzd jIng.
Bogdan, Robert. 1988. *Freak Show: Presenting Human Oddities for Amusement and Profit*. Chicago: University of Chicago Press.
Cohen, Colleen Ballerino. 1996. "Contestans in a Conteted Domain. Staging Identities in the British Virgin Island." In *Beauty Queens at the Global Stage. Gender, Contests, and Power*, edited by Colleen Ballerino Cohen, Richard Wilk and Beverly Stoeltje, 125-146. New York and London: Routledge.
Cohen, Colleen Ballerino, Richard Wilk, and Beverly Stoeltje. 1996. *Beauty Queens on the Global Stage. Gender, Contests and Power*. New York and London: Routledge.
Craig, Maxine Leeds. 2002. *Ain't I a Beauty Queen? Black Woman, Beauty, and the Politics of Race*. New York: Oxford University Press.
Davis, Lennard J. 1995. *Enforcing Normalcy. Disability, Deafness, and the Body*. London & New York: Verso.
Davis, Lennard J. (red.). 2006. *The Disability Studies Reader*. 2. Ed. New York and London: Routledge.
Garland-Thomson, Rosemarie. 1997. *Extraordinary Bodies. Figuring Physical Disability in American Culture and Literature*. New York: Columbia University Press.
Garland-Thomson, Rosemarie. 2009. *Staring. How we Look*. New York: Oxford University Press.
Garland - Thomson, Rosemarie (red.). 1996. "Introduction: From Wonder to Error – A Genealogy of Freak Discourse in Modernity." In *Freakery. Cultural Spectacles of the Extraordinary Body*, 1-23. New York and London: New York University Press.
Johnsson, Mark. 1996. "Negotiating Style and Mediating Beauty. Transvestite (Gay/Bantut) *Beauty Contests in the Southern Philippines*." In *Beauty Queens On The Global Stage*, edited by Colleen

Ballerino Choen, Richard Wilk and Beverly Stoeltje, 89-104. New York and London: Routledge.
Julén, Staffan, and Marius van Niekerk. 2010. *My Heart of Darkness*. Sweden/Germany.
Lavenda, Robert H. 1996. "'It's Not a Beauty Pageant!' Hybrid Ideology in Minnesota Community Queen Pageants." In *Beauty Queens on The Global Stage*, edited by Colleen Ballerino Cohen, Richard Wilk and Beverly stoeltje, 31-46. New York and London: Routledge.
Moskalenko, Lena. 1996. "Beauty, Women and Competition. 'Moscow Beauty 1989'." In *Beauty Queens on the Global Stage. Gender, Contests and Power*, edited by Colleen Ballerino Cohen, Richard Wilk and Beverly Stoeltje, 31-47. New York and London: Routledge.
Siebers, Tobin. 2008. *Diasbility Theory*. Ann Arbor: The University of Michigan Press.
Snider, Stefanie. 2009. "Fat Girls and Size Queens: Alternative Publications and the Visualizing of Fat and Queer Eroto-politics in Contemporary American Culture." In *The Fat Studies Reader*, edited by Esther Rothblum and Sondra Solovay, 223-230. New York and London: New York University Press.
Stoeltje, Beverly. 1996. "The Snake Charmer Queen. Ritual, Comeptition, and Signification in American Festival." In *Beauty Queens on the Global Stage. Gender, Contests and Power*, edited by Colleen Ballerino Cohen, Richard Wilk and Beverly Stoeltje, 13-31. New York and London: Routledge.
Traavik, Morten. 2009. Miss Landmine. Landmine suvivoris' fashion – Cambodia 2009. miss-landmine.org.
Wilk, Richard. 1996. "Connections and Contradictions: From Crooked Tree Chashew Queen to Miss World Belize." In *Beauty Queens on the Global Stage. Gender, Contests and Power*, edited by Colleen Ballerino Cohen, Richard Wilk and Beverly Stoeltje, 217-232. New York and London: Routledge.

Notes on Contributors

Alexandre Dessingué, Professor of literary studies and history education at the University of Stavanger (Norway). His research interests are Historical and critical literacy, Humanities in Education, Cultural memory studies, Uses of history, Literature and Cultural Theory, Discourse and textual analysis. Among his publications are: "From Collectivity to Collectiveness: Reflections (with Halbwachs and Bakhtin) on the Concept of Collective Memory" in Siobhan Kattago *The Ashgate Research Companion to Memory Studies* (London 2015), and (co-edited with Pr. Jay Winter) *Beyond Memory: Silence and the Aesthetics of Remembrance*, London 2016.

Ketil Fred Hansen, Associate professor in Social Sciences at the University of Stavanger (Norway). In his research he has been dealing with African studies especially on Chad – he is the trusted writer of the Chad-chapter in Brills Africa Yearbook. He is involved in several projects dealing with development aid – both in a contemporary and historical perspective.

Steffi Hobuß, Senior Lecturer/Associate Professor in Philosophy and Kulturwissenschaften at Leuphana University Lüneburg (Germany). She has published many papers in the fields of Philosophy of Language, Intercultural Philosophy, Memory Studies, and *Konversionen. Fremderfahrungen in ethnologischer und interkultureller Perspektive* (Amsterdam 2004, co-edited with I. Därmann and U. Lölke), *Erinnern verhandeln. Kolonialismus im kollektiven Gedächtnis Afrikas und Europas*

(Münster 2006, co-edited with U.Lölke), *Lassen und Tun* (Bielefeld 2015, co-edited with N. Tams).

Benedikt Jager, Associate professor in Nordic Literature at the University of Stavanger (Norway). His research interests are connected to the fields of Memory studies and literary criticism. His research has focused on the relationship between Norway and the GDR – especially on questions of censorship (see *Norsk litteratur bak muren. Publikasjons- og sensurhistorie fra DDR (1951-1990)*, 2014). He is member of the Norwegian Pen-Club.

Ketil Knutsen, Associate professor in History at the University of Stavanger (Norway). His fields of research are: Uses of history, history and social sciences in education, digital history, cultural theory and Kenneth Burke's dramatism. His latest publication is "A history didactic experiment: the TV series Anno in a dramatist perspective", in *Rethinking History*, 20:3 (2016): 454-468.

Nadine Siegert is Deputy Director of the Iwalewahaus at Bayreuth University (Germany). Her fields of research are: Contemporary arts, arts in Africa, arts & politics, archive, visual studies, iconography, lusophone Africa. She has published many papers about the Arts in Africa, e.g. "Gooving on broken – dancing war trauma in Kuduro?" In: L. Bischoff and S. Van de Peer (ed.): *Art and Trauma in Africa: Representations of Reconciliation in Film, Art, Music and Literature*. London: 2013.

Nora Simonhjell has a positon as Associate professor in Nordic Literature at the University of Stavanger (Norway). She is a specialist in Scandinavian contemporary literature and poetry. Her research deals mostly with issues on body and gender. In her ongoing project, Simonhjell links these subjects to questions of illness, disability and of aging and analyses the esthetics of dementia-narratives.

Jon Skarpeid has a positon as Associate professor in Religious studies at the University of Stavanger (Norway). He is a specialist in South Asian Religions and his research has dealt with the importance of music in different world religions. His PhD was a comparison of narrative structures in Indian religions and Hindustani music.

David Wagner has a positon as Associate professor in History at the University of Stavanger (Norway). He has published a book about the representation oft he banlieu in contemporary French Films. His current research interests are connected to the use of different media (film and comics) in teaching history. These issues are linked to the broader frame of digitalization of the humanities.

Kaya de Wolff is a PhD candidate at the Institute for Media Studies at the University of Tuebingen (Germany). She holds a PhD grant founded by the Hans-Boeckler-Stiftung and is a member of the junior research group "Trans-cultural public sphere and solidarization in present media cultures". Her research fields include cultural media studies, memory studies, and postcolonial studies. In the framework of her doctoral research project, she conducts an empirical study of the controversial memory discourse regarding the Herero and Nama genocide. Publications include "The politics of cosmopolitan memory from a postcolonial perspective" (forthcoming).

Cultural Studies

Gundolf S. Freyermuth
Games | Game Design | Game Studies
An Introduction
(With Contributions by André Czauderna, Nathalie Pozzi and Eric Zimmerman)

2015, 296 p., pb.
19,99 € (DE), 978-3-8376-2983-5
E-Book
PDF: 17,99 € (DE), ISBN 978-3-8394-2983-9

Suzi Mirgani
Target Markets – International Terrorism Meets Global Capitalism in the Mall

2016, 198 p., pb.
29,99 € (DE), 978-3-8376-3352-8
E-Book: available as free open access publication
ISBN 978-3-8394-3352-2

Martina Leeker, Imanuel Schipper, Timon Beyes (eds.)
Performing the Digital
Performativity and Performance Studies in Digital Cultures

2016, 304 p., pb.
29,99 € (DE), 978-3-8376-3355-9
E-Book: available as free open access publication
ISBN 978-3-8394-3355-3

All print, e-book and open access versions of the titles in our list are available in our online shop www.transcript-verlag.de/en!

Cultural Studies

geheimagentur, Martin Jörg Schäfer, Vassilis S. Tsianos (eds.)
The Art of Being Many
Towards a New Theory and Practice of Gathering

2016, 288 p., pb., numerous ill.
34,99 € (DE), 978-3-8376-3313-9
E-Book
PDF: 34,99 € (DE), ISBN 978-3-8394-3313-3

Ramón Reichert, Annika Richterich, Pablo Abend, Mathias Fuchs, Karin Wenz (eds.)
Digital Culture & Society
Vol. 1, Issue 1 – Digital Material/ism

2015, 242 p., pb.
29,99 € (DE), 978-3-8376-3153-1
E-Book
PDF: 29,99 € (DE), ISBN 978-3-8394-3153-5

Ramón Reichert, Annika Richterich, Pablo Abend, Mathias Fuchs, Karin Wenz (eds.)
Digital Culture & Society (DCS)
Vol. 2, Issue 2/2016 – Politics of Big Data

2016, 154 p., pb.
29,99 € (DE), 978-3-8376-3211-8
E-Book
PDF: 29,99 € (DE), ISBN 978-3-8394-3211-2

All print, e-book and open access versions of the titles in our list are available in our online shop www.transcript-verlag.de/en!

Printed by Printforce, United Kingdom